DATE			

WAR WITHOUT MEN

Pergamon Titles of Related Interest

Collins US–SOVIET MILITARY BALANCE

English et al. THE MECHANIZED BATTLEFIELD

Garnell GUIDED WEAPON CONTROL SYSTEMS

Lee et al. GUIDED WEAPONS

Main BEAM WEAPONRY

Mason WAR IN THE THIRD DIMENSION

Reed UNMANNED AIRCRAFT

Rodgers et al. SURVEILLANCE & TARGET ACQUISITION SYSTEMS

RUSI INTERNATIONAL WEAPON DEVELOPMENTS

RUSI MILITARY POWER SERIES

Simpkin SIMPKIN ON ARMOR

Simpkin RACE TO THE SWIFT: THOUGHTS ON 21ST CENTURY WARFARE
Volume I of the Future Warfare Series

Related Journal

(Free specimen copies available upon request)

DEFENSE ANALYSIS

WAR WITHOUT MEN

Robots on the Future Battlefield

Steven M. Shaker

and

Alan R. Wise

Volume II of the Future Warfare Series

Perry M. Smith, *General Editor*

PERGAMON-BRASSEY'S

International Defense Publishers, Inc.

(a member of the Pergamon Group)

Washington New York London Oxford
Beijing Frankfurt São Paulo Sydney Tokyo Toronto

U.S.A. (Editorial)	Pergamon-Brassey's International Defense Publishers, 8000 Westpark Drive, Fourth Floor, McLean, Virginia 22102, U.S.A.
(Orders)	Pergamon Press, Maxwell House, Fairview Park, Elmsford, New York 10523, U.S.A.
U.K. (Editorial)	Brassey's Defence Publishers, 24 Gray's Inn Road, London WC1X 8HR
(Orders)	Brassey's Defence Publishers, Headington Hill Hall, Oxford OX3 0BW, England
PEOPLE'S REPUBLIC OF CHINA	Pergamon Press, Room 4037, Qianmen Hotel, Beijing, People's Republic of China
FEDERAL REPUBLIC OF GERMANY	Pergamon Press, Hammerweg 6, D-6242 Kronberg, Federal Republic of Germany
BRAZIL	Pergamon Editora, Rua Eça de Queiros, 346, CEP 04011, Paraiso, São Paulo, Brazil
AUSTRALIA	Pergamon-Brassey's Defence Publishers, P.O. Box 544, Potts Point, N.S.W. 2011, Australia
JAPAN	Pergamon Press, 8th Floor, Matsuoka Central Building, 1–7–1 Nishishinjuku, Shinjuku-ku, Tokyo 160, Japan
CANADA	Pergamon Press Canada, Suite No. 271, 253 College Street, Toronto, Ontario, Canada M5T 1R5

First edition 1988

Library of Congress Cataloging in Publication Data

Shaker, Steven M.
War without men.
Bibliography: p.
Includes index.
1. Vehicles, Military. 2. Vehicles, Remotely piloted. I. Wise, Alan R. II. Title.
UG615.S53 1987 623.74 86–25401

British Library Cataloging in Publication Data

Shaker, Steven M.
War without men: robots on the future battlefield.
1. Warfare. Applications of robots
I. Title II. Wise, Alan R.
355′.02

ISBN 0-08-034216-7

Printed in Great Britain by A. Wheaton & Co. Ltd., Exeter

Pergamon–Brassey's Future Warfare Series is dedicated to the memory of Brigadier Richard E. Simpkin, OBE, MC, soldier and scholar, whose vision of the future inspired this series.

Preface:

Pergamon–Brassey's Future Warfare Series

War will be with us for some time. Sadly we must recognize it to be one of mankind's most enduring endeavors. Our children must learn to defend themselves and so must their children. But if we prepare now, if we anticipate the nature of tomorrow's wars and begin to counter their new dangers, perhaps that future world—the world of the twenty-first century—will be a safer one than ours.

Pergamon–Brassey's has developed its Future Warfare Series for everyone concerned about tomorrow's world. National policymakers, military and civilian teachers, defense industry executives, informed citizens, and professional military personnel will all gain valuable insights through this series that will enhance their ability to meet the challenges of future warfare. Each book, written by an acknowledged expert, is intended to stimulate the reader's thinking, to raise new issues, and to initiate debate by peering into the future, boldly suggesting those factors that will shape warfare well beyond the twentieth century. Though every volume can and will stand on its own merit, our goal is to publish a series that encourages mankind to address the full spectrum of future warfare issues—now, before the future becomes the present and overwhelms the defenses of the past.

Maj. Gen. Perry M. Smith, USAF (Ret.)
General Editor

Contents

List of Illustrations

Foreword

In the second century B.C., Hero of Alexander constructed automata that were animated by water, air, and steam pressure—statues that moved, doors that opened, and birds that sang on steam. Automata (from the Greek *automatos*, "acting of itself") have fascinated people ever since. Evolving from clockwork and gear-filled devices to robots having sensors, effectors, and processors with artificial intelligence, they are no longer simple amusements. They have become instruments of life and death.

Robotic vehicles are autonomous, semiautonomous, or remotely controlled vehicles that behave "intelligently" (although not necessarily as intelligently as a person). They operate in space, in the air, and at (or under) the sea. Spacecraft, aircraft, landcraft, and seacraft are gaining intellects hinged to the computer chip. It is primarily the military that is developing robotic vehicles; the robotic combat vehicle will be operational well before the robotic automobile. Robotic vehicles will ultimately pervade the military and society in general. They will have a profound impact on the structures and processes of our institutions.

From the World War I "Kettering Bug" to today's remotely piloted vehicles (RPVs) and other robotic combat vehicles, unmanned vehicles are solutions to problems. The military faces a high attrition of expensive systems with valuable humans on-board. Personnel are becoming scarcer with changing demographics, and personnel costs are escalating rapidly. Manned systems cost up to tens of millions of dollars, even hundreds of millions, each. Unmanned vehicles can be smaller than manned—they need not carry people and life-support gear. They can be much less expensive. They can be expendable, as with kamikaze mini-attack drones, or more survivable (because they are more maneuverable and are less visible to sight and radar), as with reconnaissance RPVs. Their sophistication can make them easier to operate, and fewer, lower-skilled personnel would be needed to operate them. They are the essence of our advantage over our adversaries: *high technology* (microchips and machine intelligence). In combat they can perform reconnaissance, attack, electronic warfare, or communications functions at high or low altitudes, on various terrains or at different ocean depths. Robotic vehicles can neutralize our adversaries' quantitative advantage without resorting to the use of tactical nuclear weapons. They can help prevent a nuclear holocaust.

The robotic revolution is underway, yet there is a no clear understanding as to what it will mean. It is a "revolution" because it is a technological discontinuity in man's history, a fundamental change like the Industrial Revolution. Now artificial minds will be added to artificial muscles. The first generation of combat robots, mainly teleoperated (remotely controlled) devices, will be primitive but useful and economically feasible. Their successors will be orders of magnitude better. The Department of Defense is spending exponentially increasing amounts on research and development of robotic combat vehicles. Aside from a few organizational and operational analyses performed for the introduction of remotely piloted vehicles into the armed services, almost nothing has been done to examine the tactical, strategic, organizational, political, and cultural impacts that might arise from having robotic air, land, and sea systems pervade the military.

Initially, the impacts are likely to be simple: having more accurate, flexible, cheaper, and reliable weapons. After a time the organization, composition, and structure of forces will likely change. There might be smaller units without separate air, land and sea services; units might be characterized by some other scheme, such as teleoperated (human-in-the-loop) robotic units, autonomous robotic units (although humans would presumably give the orders), and augmented human units (perhaps humans in robotic exoskeletons, or humans accompanied by robotic associates). Changes in tactics might include more dynamic maneuvers (the offensive defense), made possible by the fearlessness and expendability of robots and their ability to share information instantly (in such a way that humans would require extrasensory perception to duplicate) through communications linked directly to their sensors and machine intelligence. New recruiting policies and training programs would be needed to reflect the changes in personnel engendered by robotic combat vehicles. Physical ability would no longer be a requirement for effective military service. Women and older men could operate teleoperated systems, engaged in combat, from positions of safety halfway around the world. New skills will be needed, and the terms of a service career will be different.

Further in the future, the effects of combat robotics might include the realization of the Conventional Defense Initiative (i.e., the replacement of nuclear weapons and the doctrine of Mutual Assured Destruction with nonnuclear weapons without losing the deterrent effect). This implies a blurring in the distinction between tactical and strategic warfare and a potential parity of defense and offense. Along with changes in long-range military intervention in Third World countries, there would be fewer political repercussions with only robots risked in battle and not the children of voters. It may come to pass that machines will fight other machines, exhausting national wealth but sparing most people—a kind of potlatch warfare. The effects of combat robotics, although unclear now, will certainly be profound. Consider the RPV, now becoming operational in the Army, Navy, and Marine Corps. What will be the effects of having high-quality, real-time video available, from a distant battle, to anyone in the chain of command, including the president? This could lead to greater effectiveness because of the availability of more timely, accurate information; or it could lead to disaster because of kibitzing up and down the chain of command.

The computer chip is evolving beyond the point where it has the equivalent of the neuron density of the human brain. The essence of man is his brain; the essence of the robotic vehicle will be its brain. Progress will come in mechanical systems (arms, end effectors, locomotion by wheels, tracks, legs, and propellers); sensors (vision, infrared, acoustic, millimeter wave, tactile, and smell); processors (faster computers having more memory, better software for expert systems, natural languages, vision, and general computation); and control systems (on-board and man/machine interfaces). The Industrial Revolution lasted more than a century; machines replaced muscle. The robot revolution will also last more than a century; machines will replace brain matter as well as muscle. This has already started, and cannot be stopped. Society's institutions (military and civilian government, industry and commerce, labor, academe, organized religious institutions, and the family unit) must control and accommodate this revolution. It must be humane.

Steve Shaker is a prolific chronicler of combat robotics. In this, the world's first comprehensive book on the subject, he and Alan Wise successfully meet a major challenge: to survey the entire spectrum of combat robotics—from teleoperated to autonomous systems, from space to air to land and sea units, from the past to the present, from the present to the future. It is difficult to cover an entire technology that is changing from week to week, but they give the reader an accurate, comprehensive, and fascinating account of the coming robotics revolution.

Robert Finkelstein
Washington, DC

Acknowledgments

Our utmost thanks go to our families, who have been very instrumental in supporting our book. Steve's wife, Virginia Shaker, contributed many hours of word processing, as well as providing moral encouragement. Thelma Dye's assistance in typing letters requesting information from the numerous corporations and government organizations was also invaluable.

It is a real honor to have Robert Finkelstein write the foreword for this book. He is one of the finest analytical thinkers in the field of combat robotics. His monthly column in *Unmanned Systems,* put out by the Association for Unmanned Vehicle Systems (AUVS), is a real pleasure to read. AUVS (located in Washington, DC) was also extremely helpful by sharing information and recommending contacts in industry and government. Numerous government and corporate officials were very generous in offering their insight on robotic weapons, as well as furnishing information and photographs on the unmanned systems they are associated with.

Special thanks go to DeForrest Ballou, editor of *National Defense* magazine, for his advice and encouragement on writing about advanced weapons concepts, and to our production editor, Lisa Tippett.

Finally, we greatly appreciate the opportunity that Frank Margiotta, president of Pergamon-Brassey's International Defense Publishers, and other individuals there have given us to present this first book on robotic weapons systems.

1 War Without Men?

The longbow. Gunpowder. Steam-driven vessels. Rapid-fire machine guns. The tank. Airplanes. Nuclear weapons. Missiles. These inventions are merely a few of the military technologies that have changed warfare, shaken empires, rearranged societies, destroyed nation-states, changed social structures. In the twentieth century, the pace of military technological change has increased. Today we live in a world where many fear that nuclear weapons threaten the very existence of mankind and life on this planet. Around the corner, the twenty-first century approaches rapidly. Let's peer into a crystal ball, through the eyes of a teleoperated robot, and see what is going on:

The year is 2041. As we scan the battlefield through our remote television sensor, we see devastated farmlands, rubble, and wisps of smoke rising from what were once thriving villages. Suddenly, on the horizon we can see a machine that resembles a helicopter whirling across the sky. In a command post buried in the mountains of West Virginia a battle-hardened team leaps into action and energizes robotic vehicles concealed in the rubble of the once-proud German villages. The American warriors are a strange group; led by a tired, balding, portly older two-star general officer, the team consists of three somewhat ungainly teenagers (two boys and a girl) and a gaunt energetic 42-year-old woman with a doctorate in robotic science from the Massachusetts Institute of Technology.

The grandmaster general puts on his eyeglasses and squints at the video picture of the German plain. The young men and women busily flip switches while reading the sensor information from the ground- and air-based sensors and reconnaissance robots on this recent battlefield. When they rightly conclude that the whirligig is the latest Eastern Bloc reconnaissance robot, the grandmaster decides to engage. Immediately, the downy-faced young ground attack master transmits data to the hidden U.S. robot vehicles mounted with laser cannons. There is no response. The M.I.T. scientist carefully checks the communications board and alerts the combat team that an attack by ground-based Soviet missiles has wiped out the communications satellites essential to activating the robots. As she signals the space station hovering in synchronous orbit to deploy another satellite, the airborne attack master launches an attack at the command of the grandmaster. On board the huge mother ship safely orbiting 200 miles away, digitalized instructions are transmitted through the remotely piloted vehicles (RPVs) slung in the bomb bay and hung under its wings. Three RPVs are launched to attack the whirligig. As they leave the mother ship, they begin electronic jamming in response to radar signals from an Eastern Bloc space station hovering 500 miles above the atmosphere. Once the RPVs are within 100 miles of the front, they begin evasive action as they come within danger of the space station's weaponry. Laser bolts lash down at them from the sky above. Three must still be a "magic number," as two RPVs are incinerated, but the third one gets within firing range of the now retreating whirligig. Back on the airborne mother ship, the airborne tactical master maneuvers the teleoperated RPV closer to its target, moves the cross hairs over the encircled target on his scope, and transmits the death signal to the RPV. A burst of four missiles leaves the RPV and heads for the whirligig, which automatically responds with its own deadly intercept missiles in an attempt to defeat the attack. Two of the missiles pass safely through the barrage from the whirligig and strike it in the middle of the huge red star on its side. As the whirligig explodes, a signal is sent to the RPV; it turns around and heads for a pasture near the robotic mobile recovery Landrover. Unfazed by the nuclear, chemical, and biological waste that infects the charred countryside, the Rover moves to the RPV under the direction of the maintenance master in the mother ship. After picking up the RPV, the Rover crawls to the steel doors in the side of a hill and enters a decontamination chamber manned by robotic sprayers. After the contamination level on the RPV has been reduced to a manageable level, it is brought through the filtration room into the main maintenance depot, where it can be finally tended by human hands.

Several hours later, the grandmaster and his four novices assume command of a contingent of the latest version in robotic armor, the General Motors M-12 HAIG. Using this unmanned vehicle's revolutionary decoy hologram and false signature devices, the remote controllers are able to mislead the Soviet defenders to counterattack in the wrong area. The HAIGs smash through a sparsely defended front line and continue on into the Red German countryside, heading towards Berlin. A robotic "mole" intelligence interceptor that had covertly burrowed and affixed itself to a Soviet communications cable, picks up a top secret Soviet message stating that if Berlin falls, they are prepared to launch ballistic missiles at key U.S. command and control posts, including that of the robot controllers positioned deep within the West Virginia mountain. The United States responds by converting some of its space-based kinetic and laser weaponry into the offensive mode, capable of igniting Soviet oil refineries and incinerating agricultural collectives.

The grinding of teeth, cracking of knuckles, and pacing back and forth is all that can be heard in the grandmaster's command post until word is received that a truce with the Soviets has been reached and all forces are to return to their pre-established lines. The general tries to retreat his automated attackers, but the Soviets use their recently acquired jammers to disrupt the communication links between the West Virginian operators and their robotic tanks. The Russians seem unaware that by cancelling the American signals, the M12s automatically convert into their autonomous mode. The armored robots continue on towards Berlin, as they had been pre-programmed to perform. The general explains the situation to his superiors. Through the hot line, the president tries to convince the Soviet leader that the robots are not responding to commands and that the only way to bring them under control is for the USSR to halt all jamming. The KGB and military leadership, however, believe it to be a ruse and Soviet signal interference continues. As the ICBMs are being readied, and the U.S. orbital robot stations assume orbits over the Soviet heartland, the general longs for the days when wars were won and lost by the actions of people on the battlefield.

Is this how wars will be fought in the twenty-first century? Can we know what a future battlefield will look like? Will the "warriors" be tired old "grandmasters" able to teach and lead young virtuosos and highly trained scientists? Will societies be shaken and charred? Will military leadership evolve in very different directions? Will tactics and strategy become those of complex distant computer games? Obviously, no one can answer these questions with any certainty, nor accurately predict if some warfare in the twenty-first century will be waged by robots and without humans. Evidence exists, however, that enables us to forecast that there will be robots on the future battlefield and that they will play an increasingly powerful role in mankind's calculations. This book provides the first comprehensive overview of developments in military robotics, and is intended to challenge citizens, policymakers, elected officials, and military leaders to recognize the next major military technological trend that must rapidly impact upon our thinking. There will be war without men eventually, and unless we human beings control that war, it might develop in ways that could prove very costly to our descendants.

BUREAUCRATIC BATTLE

Our examination of the massive research and development of military robotics does permit us to be certain that one of the biggest battles the military will participate in during the next century will pitch men against robots. Although there will be some mortal conflict between metal and flesh, the primary struggle will be a bureaucratic one within the Pentagon and other nations' military establishments. This fight will determine whether autonomous-operating robots will function on the battlefield with minimal human participation, or whether man will retain the dominant role as warrior. Just as in industrial production, the trend towards automation and the use of robotics has begun, and it is picking up momentum. More and more professional soldiers are moving away from combat duties to serve at computer terminals. In some instances, traditional roles performed by manned systems have already been replaced by automatic mechanical devices. The technology for substituting for other human-driven systems with robotics is rapidly materializing.

It is a natural tendency of many military leaders to fight the next war with proven tactics and strategies and to rely on proven military hardware. Professional soldiers are often skeptical of conceptual weapon systems dreamed up by civilian engineering "whiz kids," who, they believe, do not understand real warfare conditions. They especially dislike weapon systems that need little direct control. There is even something unromantic about warfare being fought by objects with no emotional commitment to their cause. The end result is that rapidly developing weapon technologies threaten the roles of the individuals within the bureaucracies who are managing those technologies' development. Therefore, the sociological obstacles to robotic warfare may be more difficult to overcome than the technological ones. The U.S. military's slowness in fielding remotely piloted vehicles (RPVs) and other "smart weaponry," even though much of the technology is already at hand, may be attributed to this mind set.[1] A U.S. General Accounting Office report to Congress substantiates this by stating that the Department of Defense has exhibited a reluctance to adequately fund or utilize RPV technology. It also found that there is a need to assure the widest utilization of this technology where its use can save lives and money.[2]

EVOLUTIONARY STEPS

Robotic weapon systems can best be defined as autonomous, semiautonomous and teleoperated artificial systems or vehicles that perform military missions thought to be appropriate for human beings.[3] This definition is somewhat flexible, and it may still be open to subjective interpretation as to whether a particular system should come under the robot heading. In general, the weapon systems covered by this book include remotely piloted airborne vehicles, unmanned naval vessels and submersibles, unmanned spacecraft, and robotic ground vehicles.

The evolution of unmanned military vehicles began as early as World War I, when miniature airplanes packed with explosives were hurled at their targets. The likelihood of a direct hit was based in part on the proficiency of the launch crew, but more importantly, on pure luck. After estimating the target's distance and allowing for wind drafts, the land crew would point the unmanned biplane in the proper direction. The engine would be set to run for the estimated time it took the aircraft to reach the target. The aircraft's performance was slightly improved with the addition of a gyroscope and aneroid barometer to control direction and altitude. Such systems were very crude and ineffective and, thus, were seldom used. Nonetheless, they were the ancestors of latter-day guided missiles and remotely piloted vehicles.

Unmanned vehicles took another evolutionary jump with the introduction of radio-controlled aircraft at the end of World War I. By the 1930s, many well-to-do hobbyists throughout the world were flying radio-controlled airplanes. The first automatic pilots were developed by the British in the late 1920s, and were effectively put to use in the German V-1 flying bombs during World War II. Also, in the 1940s, radio-controlled flying drones were used as aerial targets, and remote-controlled minitanks were employed to detonate mines.

Mobile target technology steadily improved through the 1950s. At the same time, elementary cruise missiles were developed, and although their accuracy was questionable, they had the ability to span continents to reach their targets. Cruise missiles were eventually rejected by the U.S. Air Force, which preferred to put its money into manned bombers. The advent of ballistic missiles replaced many of the proposed missions for the cruise missile, thus furthering its demise.

The use of remotely piloted vehicles blossomed during the Vietnam War, when manned reconnaissance aircraft found it too hazardous to approach heavily defended surface-to-air missile (SAM) sites. Along with reconnaissance and intelligence gathering, RPVs were used as electronic harassment decoys and even to drop propaganda leaflets. The space program of the 1960s and early 1970s, which included technology used in the unmanned spacecraft landings on the Moon, Mars, and Venus, and the probe fly-bys of the outer planets, also influenced military robotic programs. Cruise missiles were brought back into operation in the late 1970s. Greatly improved programmable technology and the ever-increasing difficulty facing manned bombers for deep penetration into enemy territory made the adoption of cruise missiles more acceptable to the "top brass." Current cruise missiles can travel over 1,000 miles; by using on-board computers and sensors, they can follow the terrain to their intended target. Nevertheless, the military service bureaucracies, dominated by aviators, ship captains, and tank commanders, are still somewhat reluctant to adopt systems that could place them in the less glorious positions behind the front.

The developments in Lebanon in 1982 demonstrated the necessity for the widespread adoption of unmanned military vehicles. Much of the credit for the Israeli destruction of some seventy-nine Syrian aircraft and nineteen SAM sites while only suffering the loss of one aircraft was attributed to their use of RPVs. The Israelis ingeniously used their RPVs to obtain Syrian SAM radar signatures. The information was used to jam and confuse the Syrian operators and then to destroy the SAM sites with radiation-homing missiles.[4]

Shortly following the Israeli successes, the U.S. Navy lost three aircraft to Syrian antiaircraft batteries during a retaliatory strike against Syrian positions in Lebanon's Bekaa Valley. The cost, including the loss of life, aircraft, and political dignity suffered by the United States, was far greater than any benefits obtained by the attack. The navy had to turn to the battleship U.S.S. *New Jersey* to pound the Syrian positions with its big 16-in. guns. The inability to get observers into Lebanon to provide accurate spotting for the battleship downgraded its effectiveness. These Lebanese experiences helped to convince U.S. military skeptics of the need to introduce advanced RPV systems to perform some of the missions done by manned aircraft.[5]

Current robotic weapon systems reflect either remote-controlled or programmable technology.[6] The simple remote-controlled systems are giving way to programmable vehicles capable of taking action in response to situations anticipated by the programmers. Recent accomplishments in supercomputer architecture, artificial intelligence (AI), robotics, and sensor and vision research are incrementally moving unmanned vehicles into their ultimate evolutionary state, where they can reason on their own as to the best course of action to achieve a goal. The Department of Defense's Advanced Research Projects Agency (DARPA) has embarked on an ambitious plan to produce within the next few years an autonomous land vehicle (ALV) that can independently move around impassable objects and traverse over rough terrain at speeds up to 25 mph. The opportunities for weapon superiority afforded by these technologies, as well as the increasingly dangerous battlefield environment, may eventually relegate man to the role of behind-the-scenes strategist, leaving machines to perform the actual fighting. If current trends continue, it is not a question of whether this will happen, but rather how long it will take.

TO BE OR NOT TO BE AUTONOMOUS

Lumping remote-controlled robots with programmable and autonomous robots is like grouping together slide rules, calculators, and supercomputers. They all were designed for similar purposes, although the level of sophistication from which each attacks the problem is quite different.

Remote-controlled robots are maneuvered by signals conveyed through a tethered-cable attachment, or by a radio link. Each movement of the robot corresponds to the commands given by the human operator. Current technology has made the controls easier to handle, and in some systems the controls have come to resemble the same type of joysticks used in video games. The operator may receive on a monitor images that are obtained from a TV camera, infrared camera, or other sensor mounted on the robot. The human operator, although he might be quite distant, still serves as the robot's brain. Remote-controlled robots exhibit no intellectual abilities, and on the evolutionary scale would be considered sublife.

Programmable robots, such as that in the Tomahawk cruise missile, already house their own computers containing software programmed with maps and complicated sets of directions and maneuvers. Although humans may receive some feedback by which to monitor the robot's progress, the automated mechanism is no longer dependent on real-time instructions from an operator and can function independently upon launch. Programmable robots emulate the behavior patterns of the most primitive forms of life, whose actions are instinctive and have been programmed by their genetic code. They are unable to stray from their programmed instructions. Autonomous robots, however, will be equipped with sophisticated computers and sensor systems that will allow the robot to make its own decisions without any direct human involvement.

Some creative thinkers are advocating that current missions ranging from intelligence-gathering to antitank roles be taken over by unmanned systems. Airborne RPV systems can already accomplish many of these tasks. Ground systems using remote-control or programmable technologies can be introduced to perform some basic combat missions in the near term. For example, battlefield roles such as mine clearing can be accomplished by remote-controlled ground robotic vehicles, and the U.S. Army plans to introduce such a system within a few years.[7] Completely autonomously operating robots utilizing their own artificial intelligence still require additional research and development; according to some experts, it may take until the year 2005 before a "smart" robotic system can be deployed on the battlefield.[8] It would be unwise to push complex missions on to autonomous robotic systems until the technology is further developed. Opponents of unmanned vehicles might put robots to the task and then dwell upon their shortcomings in order to cancel their further development.

Survival in battlefield conditions will require independently acting robots to make complex decisions based upon information they receive about their environment and external threats. The robots will have to choose among various difficult options, including what route and terrain are best to traverse to reach an objective. The unmanned system must house a computer that can collect information or input from its sensors (which is a way of learning new things), collate the data and synthesize new concepts out of previously obtained information, and then make original value judgments (which is how decisions are derived). One major challenge in obtaining such artificial intelligence for autonomous robots is to develop the capability for parallel processing. To date, machine intelligence has been accomplished primarily through sequential programming. Data are processed in a step-by-step fashion, at very high rates of speed. The human brain, however, functions by processing numerous different sequences at once, and in phased relationships. These include real-time thoughts, memories, sensations, and deducements. The major problems to be faced are how to design robotic system architecture so that the individual processors can communicate constantly with others and how to write software for effective parallel processing.[9] This is one of the primary goals for military-related AI research, and it is still at least a decade away.

FORESIGHT

Another obstacle facing the developers of autonomous robots is furnishing them with the vision and sensor technology that they need to adequately perceive their environment. In order for robots to perform activities such as traversing cross-country terrain, some AI theoreticians have estimated that it would require approximately 100 billion operations per second of computer processing power in order to perceive the required imagery, which is roughly 100 times greater than current processing capability. Parallel processing and very-high-speed integrated circuit (VHSIC) technologies may allow vision researchers to obtain such levels of processing power.

How to get the robot to see is a very difficult problem, although there are several different approaches to doing so. The gray-scale technique relies on digitized images made up of thousands of little squares, called pixels. Each square reveals a shade of gray that corresponds to the light intensity picked up by the robot's camera system. Variation in brightness allows the robot's computer to determine texture. Current gray-scale systems are limited because the computer takes too much time to process their information. Researchers want to process the data images in real time without reducing the number of pixels, because the more pixels there are, the greater the image's resolution. One approach being explored is called "windowing." Only selected groups of pixels out of the entire image are processed; they therefore do not take up computer memory or time. For example, a robot does not need to perform additional processing to discern a rock from a tree in order to decide to go around an object.

The stereo vision system approach uses two cameras to mimic the human depth perception created by the overlapping of views seen from each eye. Each eyeball sees the object from a slightly different angle, and this variation is known as a parallax. The stereo vision system plugs in the different coordinates obtained from two cameras into a complex algorithm that computes the shift due to parallax. Using this system in conjunction with a filtering technique produces depth measurements. A third approach to robotic vision is called structured light. A camera and a light source are used to determine an object's location, orientation, and surface characteristics. By knowing the distance between the camera and light source plus one angle, the distance from the robot to the object can be triangulated.

In May 1985, DARPA demonstrated the ability of its ALV prototype to follow the path of a road. Using a specially designed TV camera and advanced image-processing techniques that could process a frame of imagery every 2.4 seconds, the ALV was able to locate the edges of the road, track the center lines, and move down the road accordingly. The ALV will soon be tested in its ability to move over paved, gravel, and dirt roads at speeds in excess of 12 mph.[10]

PHASING ROBOTS ONTO THE BATTLEFIELD

A more immediate and important application for unmanned ground vehicles, although less glamorous, is that of rear-area logistic vehicle. It would be a more prudent approach for robotic advocates to seek to automate logistic missions before thrusting complex combat missions on the robot. We now have the base technology in self-navigation to replace some of the human-operated truck and manual labor involved in moving supplies. Not only can supplies be moved more efficiently and effectively through automation but soldiers can also be relieved of these duties and be redeployed to the front. Industry has already developed and installed flexible manufacturing systems where automated carts follow a predetermined path from work station to work station. Industrial robots or numerically controlled machine tools perform a particular set of operations on the objects in the cart at each work station. Flexible manufacturing systems have reduced the number of personnel in an entire factory to just a skeletal crew of a few individuals.

Although more complicated than the automation of factory logistics, the core technology for the movement of various military supplies through the use of unmanned ground systems already exists. Robotic vehicles can travel predetermined paths on roads or cross-country by following special road markers, fence posts, or other signalling devices. Manipulating mechanical appendages that resemble arms, wrists, and fingers already exist, and they have the strength and dexterity to lift and package various supplies. We now have the capability to build an unmanned mobile transport system that can load and unload supplies and move them to a predetermined destination.

There is a natural tendency for military-related technologists to pursue the most exciting applications for their inventions. Advanced battlefield roles for autonomous vehicles have merit and should be researched and developed; with time such combat missions will be handled by thinking robots. In the meantime, however, simple remote-controlled and programmable robots for combat and logistic missions should be emphasized. The technology and expertise to introduce such vehicles is basically in hand, and the lessons learned in their development can be used to support the development of autonomous battlefield robots.

The widespread introduction of autonomous robots on the field of combat will probably occur within the next 20–50 years. Upcoming generations of military officers will be as thoroughly schooled in the application of artificial intelligence and robotics as they are trained to become proficient pilots or tank or ship commanders. Leadership positions will require skills similar to that of a grandmaster chess player, in which unmanned systems will be maneuvered like pieces on a chessboard. One possible benefit may be that our senior military leaders will be able to devote more time and effort to being grand strategists and tacticians rather than having to be involved in detailed operational matters.

To the uninitiated, these predictions may seem like science fiction, but these forecasts are based upon the evolution of military robotics and the increasing use of robots in war. In the next chapter, we will trace this history and then devote four chapters to the major trends in specific robotic developments for ground vehicles, airborne systems, naval uses, and space-based vehicles. Our research uncovered an enormous amount of military robot development in many countries. This volume will carefully catalog these systems and provide a brief description of each of them. We will literally be listing hundreds of developments, some more promising than others. Chapters 3 to 6 may not be the most exciting reading, but we believe it is essential to provide this baseline reference work from which important conclusions and predictions can be drawn. This unprecedented cataloging permits us, in the last chapter, to challenge those who must think about military problems to incorporate robotic systems into their planning.

NOTES

1. Wilson, George C. "Unmanned Weapons Gain Backing: Official Stresses Saving Lives, Money." *The Washington Post,* September 5, 1985, p. A15. Wilson describes frustration that former senior DoD officials had in trying to get the military services, particularly the army, to adopt unmanned systems.

Caldwell, Harlin A. and Kennedy, Jr., Floyd D. "RPV Stepchild of Unmanned Vehicles." *National Defense,* September 1982, pp. 16–20, 31, 32. The authors criticize the leisurely approach that the U.S. military has taken in acquiring RPV systems, and the lack of innovative tactics developed to keep up with the technology.

Carus, W. Seth. "U.S. Procurement of Israeli Defense Goods and Services." *AIPAC Papers on U.S.–Israel Relations 1984.* American Israel Public Affairs Committee, Washington, DC, 1984, pp. 16–17. Carus relates how both the U.S. Department of Defense's Advanced Research Projects Agency and Israel initiated significant programs to develop mini-RPVs following the 1973 Arab–Israeli War. By 1978, less than four years later, Israel had fielded two mini-RPVs (the Scout and the Mastiff). Although the U.S. Army has spent more than $350 million on their Aquila mini-RPV program, it was not put in service until 1986, more than 12 years after the program's inception.

2. *DOD's Use Of Remotely Piloted Vehicle Technology Offers Opportunities For Saving Lives And Dollars*, Report to Congress, NASAD 81–20. U.S. General Accounting Office, Washington, DC, April 3, 1984. This report lends credence to the assertion that the military has neglected robotic vehicle usage even though the technology has already proved itself in many missions.
3. Asimov, Isaac and Frenkel, Karen A. *Robots: Machines in Man's Image.* Harmony Books, New York, 1985, p. 12. This book describes robots as "artificial devices (often pictured in at least vaguely human form) that will perform functions ordinarily thought to be appropriate for human beings."

   Finkelstein, Robert. "Opinion." *Unmanned Systems*, Fall, 1983, p. 5. Finkelstein compares the official Japanese definition of robot to that of the Association of Unmanned Systems for unmanned systems, and indicates that although they are very similar, the term "robot vehicles" has a more positive connotation than that of "unmanned systems." The Japanese definition is "A mechanical system that has flexible motion functions analogous to those of living organisms, or that combines such motion functions with intelligence functions (judgment, recognition, adaption, or learning), and that acts in response to the human will." The Association for Unmanned Vehicle Systems states on the contents page of each issue of *Unmanned Systems* that "While there is no official definition of UVs, they may be thought of as autonomous or semi-autonomous vehicles or weapons which perform various functions as if a person were aboard."
4. The editorial "Beating Swords into Lemons," which appeared in the *New York Times,* November 27, 1984, p. A30, speaks to the success of the Israeli use of RPVs in Lebanon and criticizes the U.S. Army's Aquila program for "gold-plating."
5. Kennedy, Jr., Floyd D. "Unmanned Vehicles at Sea," in his "Sea Services" column in *National Defense,* November 1984, p. 12. Kennedy describes the downing of the American naval aircraft by Syrian antiaircraft batteries, which makes the case that certain missions would be better handled by unmanned aircraft.
6. Meystel, A. "IMAS: Evolution of the Unmanned Vehicle Systems." *Unmanned Systems,* Fall 1983. Meystel compares remote control with totally autonomous technology. He points out that although remote control presumes involvement and access to the information storages containing any relevant knowledge, autonomous devices have to perform all decision-making and control procedures on-board in real time.
7. Hudgins, Doris. "Robat Prototypes are Taking Shape Here.' *TACOM Report,* May 30, 1985, pp. 4–5. Hudgins provides details on the Army Tank–Automotive

Command's Robat vehicle, including technical description, planned missions, and acquisition schedule.

8. Gavin, Franc. "Robots Go to War." *International Combat Arms,* July 1985, pp. 18, 21. Describes three stages to the development of the Autonomous Land Vehicle, consisting of near-term, midterm, and far-term objectives. Far-term objectives involve the development of a "true" autonomous intelligence and are expected to take 20 years.
9. Gavin, "Robots Go to War," p. 18. Gavin discusses sequential versus parallel processing and its relationship to military robotics.

   Lemonick, Michael D. "Supercomputers: Will They Surpass the Brain?" *Science Digest,* October 1985, pp. 56, 100. Lemonick relates how researchers at Columbia University have constructed a prototype parallel processing machine that breaks down a problem and works at several pieces of it simultaneously, rather than doing one computation at a time.
10. "DARPA's Pilot's Associate Program Provides Development Challenges." *Aviation Week & Space Technology,* February 17, 1986, pp. 45–52. This article details the ALV's accomplishments and projected milestones.

2 The Evolution of Military Robotic Systems

MYTHS, METAL MEN, AND WALKING MACHINES

Early History

Inventiveness has typically been preceded by imagination, from mankind's earliest discoveries to our most recent engineering marvels. Throughout history, myths and stories have excited and motivated people to match the feats and accomplishments spoken of by the storytellers. Ancient metallurgists received respect for their ability to craft weapons and statues, which often depicted warriors. Appreciation for the strength and durability of copper, bronze, and eventually steel artifacts somehow became entwined in man's creative speculation about himself. Myths were developed of human-like creatures made from metal who, as the "supermen" of their day, were invulnerable to the weapons of the mortals. The ancient Greek god of fire, Hephaestus, was said to have fashioned and brought to life a bronze statue for King Minos of Crete. This archetypal robot ceaselessly circled the island, protecting it from attack. In a similar fashion, the bronze statue of Talos was created by Daedalus and animated to guard the sacred island of Thera. This metal monster caused much grief for Jason and his Argonauts.

Artificial men also received a great deal of attention during the Middle Ages. Legends concerning "iron men" and metal sculpted heads that spoke of coming events were quite common. Pope Sylvester II was described in some accounts as creating a "talking head" that could predict the future. He was actually quite an inventive genius who constructed a sophisticated clock relying on a system of weights and gravity to turn wheels at regular time intervals, which moved a hand around a dial. Roger Bacon, the British philosopher and inventor of the thirteenth century, also experimented with metal parts resembling human anatomy. There is a story of a sixteenth century rabbi named Loew who, in order to protect the Jews against false accusations of the ritual murder of Christian children, created an artificial man called a *Golem* to guard his people from reprisals.[1]

By the eighteenth century, as a by-product of technology gained from clock-making miniaturization, very sophisticated animated dolls were crafted. Swiss clockmaker Pierre Jaquet-Droz developed human-like machines that could play several notes on musical instruments. His automatic scribe invention featured a boy doll who could write any desired text of forty characters or less. One of the more humorous inventions of the day was Jacques de Vaucanson's mechanical duck, which could quack, drink water, eat grain, and void. Vaucanson used the duck on tours to raise money in order to further his experiments aimed at creating life artificially. After three years of effort, Vaucanson gave up and went on to develop more pragmatic inventions, including the automatic weaving loom. German Baron Wolfgang von Kempelen constructed a life-size chess player of a mechanical Turkish man who sat presiding over a chess board. The "Turk" was able to engage in a game of chess, which some suggested was a result of trickery. Von Kempelen was accused of concealing a midget within the chess player.

Twentieth Century

Early in the twentieth century the idea of "metal men" became firmly entrenched in our literature and motion pictures. The play *R.U.R.* (for Rossum's Universal Robots), composed by the Czech writer Karel Capek, was credited with coining the term robot. Robot was derived from the Czech word *robota* which means serf or forced labor. The plot centers around android machines that were designed to perform mankind's tedious work in order to make a better life for all humanity. The robots could think for themselves, but were devoid of emotion. They eventually rebelled and wiped out humanity to form a new civilization based on intelligent robots.[2]

Since the 1920s, robots have become a mainstay of science fiction. One noted author, Isaac Asimov, in his short story "Runaround," which appeared in the March 1942 issue of *Astounding Magazine*, delineated three famous rules to guide intelligent robot behavior. Asimov's Three Laws of Robotics are as follows:

1. A robot may not injure a human being, or through inaction, allow a human being to come to harm.
2. A robot must obey the orders given by human beings except where such orders would conflict with the first law.
3. A robot must protect its own existence as long as such protection does not conflict with the first or second law.[3]

The technology and trends revealed in this book imply that Asimov's laws will remain as fiction.

The dawn of the real world of robotic ground systems began in 1918 when Mr. E. E. Wichersham, an engineer with the Caterpillar Tractor Company, designed and developed a remotely controlled demolitions carrier. This vehicle, called the Land Torpedo, was battery powered and followed directional signals relayed by cable. It was never employed in combat.

In the early 1920s, engineers at the U.S. Naval Research Laboratory (NRL) in Washington, DC, built a vehicle named the Electric Dog. This early pioneer of remote-controlled ground vehicles was a three-wheel cart improvised from a child's tricycle and was driven by small series of motors supplied from a storage battery. Its radio-control system was an enhanced version of that used to operate the German unmanned torpedo boats of World War I. The control switch for operating the vehicle consisted of a small vertical shaft that resembled the control stick of 1920s vintage airplanes. The four circuits controlling the cart were connected for forward, reverse, and right and left turn. The Electric Dog provided quite a spectacle for the employees of NRL as it wandered slowly about in the driveways at the navy facility. This vehicle and its technology demonstrated the successful simultaneous and independent operation of control circuits. The navy used this technology to serve as a test bed for the remote control of aircraft (remotely piloted vehicles) and target ships rather than as a program to develop unmanned ground systems.

It is somewhat ironic that the Electric Dog was based on World War I German remote-control naval ship technology, and it was the Germans who picked up where the United States left off by producing the first operational remote-controlled ground vehicle during World War II. Mine clearing was an area of concern for the German army at the start of World War II and the idea of destroying mines with cheap, expendable remote-controlled vehicles became their engineers' favored approach. In 1939, the Borgward Company of Bremen began development of a small 8,068-lb (3660-kg) vehicle named the B1V Demolition Vehicle. It used the chassis of a tracked load carrier and had a large explosive charge container hooked to its front. A man would drive the vehicle until he considered driving to be unsafe, at which point he dismounted and the B1V then resumed its mission through radio remote control. If everything worked right, the vehicle, upon reaching its target, would release a charge with a delay device, and then backed away. The charge and subsequent detonation of the mine was not supposed to go off until the vehicle had reached a safe distance. In actual usage, however, the delay mechanism often failed, which occasionally resulted in the destruction of the B1V by the detonated mine. Altogether, 500 Borgward B1V Demolition Vehicles were built throughout the war.

The German army also saw advantages in using remote-controlled vehicles as offensive weapons. The Goliath Demolition Vehicle (Figure 2.1)

Figure 2.1. The Goliath Demolition Vehicle. (Courtesy of National Museums of Canada, Ottawa.)

was originally conceived of as an inexpensive remote-controlled vehicle that could carry a demolition charge to a target such as a pillbox, and then be sacrificed upon its detonation. Its secondary mission was as a mine-clearance vehicle, where it would set off mines by the blast overpressure created by its charge's explosion. This small tracked unit was originally propelled by electric motors and was also built by the Borgward Company. Its statistics:

- 4 ft, 11 in. (1.50 m) long;
- 2 ft, 9 in. (0.85 m) wide;
- 1 ft, 10 in. (56 cm) in height; and

The Goliath was constructed of pressed metal, with electric motors powered by batteries carried in sponsons, around which ran the tracks. A drum situated at the rear of the vehicle dispensed 0.9 mi (1.5 km) of a three-cored cable. Two strands were used to transmit steering signals; a third strand transmitted the firing signal. The operator's control over the Goliath was maintained through a small hand-held box that had switches and batteries to regulate the relays in the vehicle. This vehicle could deliver a 132-lb (60-kg) charge of explosive. Over 2,500 of the electric-powered Goliath vehicles were constructed and the Germans used them effectively in both demolition and mine-clearing tasks. The range and power of the electric-run vehicle, however, was found to be quite limited in cold weather, so a second version powered by a 703-cc twin-cylinder motorcycle engine was developed. Over 4,500 of the petrol-run Goliaths were built by the Zundapp Company of Nuremberg. This version was slightly larger than the electric model and could deliver up to a 220-lb (100-kg) charge. By 1944, numbers of Goliaths were deployed on the Atlantic coast as a defensive antitank measure.

In 1940, the British conceived of the first military-related walking machine. This was a result of the realization that although aircraft could fly swifter than birds, and naval vessels could swim faster than fish, no ground vehicle had ever matched the speed and agility with which a horse could traverse broken terrain or a mule could climb a steep hill. Although wheeled vehicles operate smoothly over level and hard surfaces, they are quite inefficient in very soft soil and in rugged land where there are projections and depressions greater than the diameter of the wheel. Tracked vehicles offer greater ability to cross natural terrain, but their maneuverability on uneven ground is still quite poor. The British firm W. H. Allen & Company, under the direction of A. C. Hutchinson and F. S. Smith, therefore designed a four-legged, thousand-ton walking tank whose leg suspension function allowed it to operate effectively on uneven ground. They built a model that featured a rolling thigh joint controlled by flexible cables connected to a console. An operator used his feet on pedals to control the two hind legs and his hands to manipulate handles to maneuver the forelegs. Although the model successfully climbed over a pile of books, the U.K. War Department cancelled funding in order to emphasize higher-priority programs offering near-term solutions to current needs.

By the late 1950s and early 1960s, the U.S. Army had examined several legged-vehicle concepts, including a bipedal robot. General Electric (GE) built the walking truck Quadroped (Figure 2.2); funded by the U.S. Army Tank–Automotive Command, which was the largest such endeavor. Although not technically a robot in itself because the vehicle was driven by an operator, it did incorporate extensive robotics-related technology and deserves mentioning. The walking machine utilized cybernetic anthropomorphous machine systems (CAMS) technology, in which the machine's motions are designed to mimic the human operator's movements. The operator was situated within the vehicle's cab and underneath were located four mechanical legs. The right and left controls for the operator's hands moved the Quadroped's right and left front legs, while the hind legs were maneuvered by the operator's foot controls. Altogether there were 18 separate inputs by the operator to manipulate the machine. Hydraulic actuators, driven by high-pressure oils, were located on each leg and powered the machine's movements. The operator was capable of driving the GE prototype while being blindfolded.

The 3,000-lb (1,350-kg) test vehicle revealed that it could climb obstacles, lift a jeep out of a mud hole, and load and unload a 500-lb (225-kg) crate of ammunition. The man/machine interface, however, became excessively involved. The walking truck's operation demanded continuous movement of the operator, whose physical and mental stamina were found to diminish

Figure 2.2. Research prototype of the Quadroped Walking Truck. (Courtesy of General Electric.)

too quickly. Generally, within fifteen minutes the individual was too exhausted to continue operating the vehicle. This serious drawback led to the cancellation of the program.

While the GE program was ending in the late 1960s, researchers at Ohio

State University (OSU), under the direction of Dr. Robert McGhee, began their own walking-machine effort (Figure 2.3). This program used computers to coordinate leg movements. The OSU scientists based their Adaptive Suspension Vehicle (ASV) in part on what they learned from Mother Nature. Studies of insects revealed that six legs was an efficient number as long as three legs were always on the ground for stabilization. Through the use of a vertical gyroscope and pendulums, the ASV adjusted its six-legged hexapod to keep the body level when traveling over irregular terrain. Consultations with entomologists also revealed that spiders, such as daddy longlegs, used their longest pair of legs as feelers in which to judge their gait. The spider's navigation was aided by sweeping the ground ahead of him with these legs. The hexapod mimicked nature by incorporating force sensors into the ASV's feet to assist in conforming to the terrain. An on-board computer was used to choose the proper footholds; it included a laser-scanning system and incorporated algorithms to coordinate gaits. The OSU program, which was in part funded by the Defense Advanced Research Projects Agency, was generally considered to have been quite a success. The driver of the 10-ft (3.08-m), 5,000-lb (2,250-kg) ASV prototype was able to step over 6-ft (1.85-m) ditches and climb over 6-ft walls. As a result of this work, which was completed in the mid-1980s, DARPA initiated a new project that combined the OSU legged vehicle with the autonomous vehicle and artificial intelligence research being conducted by Martin Marietta Aerospace Company.

FROM BUG TO BEIRUT

Early History

The Wright brothers' flight in 1903 is often considered to be the "dawn of aviation," yet the movement of objects through the atmosphere in the form of kites, gliders, and lighter-than-air vehicles such as balloons and airships had been described centuries earlier, in both myths and historical accounts. An ancient Hindu poem called the *Mahabhasata of Krishna-Dwaipayana-Vyasa* describes how Krishna's enemies induced demons to build a manless winged chariot from which missiles were hurled down upon Krishna's followers.[4] In Greek mythology, a golden owl was given by Zeus to Perseus; it aided in the rescue of Andromeda from a Kraken sea monster. Perhaps inspired by this tale, Archytas of Tareton, a friend of Plato, was said to have fashioned a mechanical bird that was suspended from the end of a bar and was revolved by a propulsion system that emitted a jet of compressed air. Ancient Chinese writings describe a warlord of approximately two thousand years ago who utilized large kites to carry explosives over a walled city and fortress. He was able to attack his adversaries while keeping his own troops out of reach of the defender's weapons. In the thirteenth century, Johann

Figure 2.3. The Adaptive Suspension Vehicle with designer Robert McGhee standing in front. (Courtesy of Ohio State University.)

Muller, a German astronomer, was credited with inventing a metal eagle that could fly. More recent historical records reveal that the French scholar Charles Rogier designed an aerial balloon system in 1818 from which rockets could be fired at an enemy by using a delayed-action device. Rogier recommended that a slow-burning fuse be installed on the balloon, thus destroying any evidence that it ever existed. This would help prevent discovery of how the weapon was constructed, as well as add some mystery to the whole operation, thus playing on the enemy's fears. During the 1890s, U.S. Army researchers based at Madison Barracks, New York, experimented with an aerial photography system that hung a camera from a large kite.[5]

World War I

Although manned-aircraft development was still in its infancy during World War I, military weapons planners and researchers soon realized that there were definite advantages in having pilotless aircraft around as well. The heavy British flyer casualties at the hands of the German Fokker monoplanes early in the war, and the inability to intercept Zeppelin airships, led to a research program at the Ordnance College of Woolwich in remotely controlled pilotless aircraft that were designed to glide into the target and explode on impact. Although the program was officially described as an effort to develop an aerial target (AT), its head, Professor A. M. Low, was secretly asked to produce an aircraft that could serve as both an interceptor and ground-attack weapon. The first Low prototype used components from existing aircraft in order to save on developmental costs. Consequently, the airframe was out of scale with the rest of the aircraft, which resulted in numerous problems such as the engine causing interference with the primitive radio-control system.

The Royal Aircraft Factory, borrowing from Low's equipment and experience, began a parallel program with several aircraft manufacturers. Companies such as Sopwith Aviation Company and DeHavilland built AT bodies to fit around the first light-weight (35-hp) expendable engine, which was produced by A. B. C. Motors Ltd. The Sopwith AT, a biplane with a wingspan of 14-ft (4.27-m), could carry a 50-lb (22.5-kg) warhead in its nose. A control box containing relays, a receiver, and a key system was located behind the fuel and oil tanks, and the aerials were wrapped around the rear fuselage and wing tips. The Sopwith program was abandoned after the prototype was damaged. During a test flight of the DeHavilland model for a gathering of important Allied dignitaries, the AT went astray and dove upon the guests, who scattered in every direction. Although none of the officials were injured, the incident brought irreparable harm to the AT program. Several other British AT efforts also experienced mishaps, and the "War to End All Wars" ended without the deployment of any RPV systems for the Royal Navy or Air Force.

The United States also conducted significant "flying-bomb" research during World War I. Dr. Elmer A. Sperry developed a gyro-stabilization unit for aircraft in 1915. This device was the heart of the autopilot, which successfully demonstrated "hands-off" flying, relieving the pilot of some of his more arduous tasks. In 1917, the U.S. Navy gave Sperry $50,000 to carry on experimental work with what was then termed "aerial torpedoes."[6] Sperry's son, Lawrence, headed the effort, which centered around automating the U.S. Navy N-9 Curtis seaplane. The N-9s were installed with automatic stabilizing, steering, and distance gear. Pilots, however, would actually perform the takeoffs and then activate the automatic systems in-flight.

The navy initiated another flying-bomb program, incorporating a Curtis airframe powered by a Ford 40-hp engine and automated by Sperry. This was a completely pilotless aircraft weighing 500 lb (225 kg) empty, which could travel at speeds approaching 90 mph (144 kph) for a distance of 50 mi (80 km). The Curtis flying bomb was not radio controlled, but was flown by presetting the gyroscope for direction and the aneroid barometer for altitude. Once the estimated distance for the target was reached, the engine was stopped, and a mechanical device removed the bolts holding the wings in place. The fuselage and explosive would then drop on the target. The Curtis flying bomb was launched on March 6, 1918, and flew a prescribed course of some 1,000 yd (900 m), thus making this the first successful flight of a "robot" aircraft.

Not to be outdone by the navy, the U.S. Army also started an unmanned aircraft program. Charles F. Kettering of Dayton, Ohio, led this effort to construct an aircraft that was cheap to build and easy to ship, assemble, and launch in the field. Kettering enlisted the aid of Orville Wright as aeronautical consultant, and C. H. Vills of the Ford Motor Company as engine consultant. The final design, known as the Kettering Bug, featured a conventional biplane with a 15-ft (4.62-m) wingspan, a gross weight of 530 lb (238.5 kg), and a 37-hp engine (Figure 2.4). It was designed to carry a 180-lb (82-kg) bomb a distance of 40 mi (64 km) at 55 mph (88 kph). The U.S. Army Air Service began flight tests of the Kettering Bug in September 1918. The first few flights were plagued with mishaps, but the fourth flight on October 22, 1918, was a complete success. The vehicle's controls were stopped at a set distance and it went into a nosedive, crashing right on target. The war ended less than a month after the fourth flight.

At the time the armistice was signed, the U.S. government was also testing the Giles self-propelled projectile built by inventor Lloyd Giles (Figure 2.5), which was comparable to the Kettering Bug in capability. The U.S. Army continued to test the Bug and in late 1918, fourteen flights were attempted, of which only four were considered successful. Although these statistics could hardly be considered startling, the military was impressed

Figure 2.4. Kettering Bugs in line for takeoff. (Courtesy of Air Force Museum archives, Dayton.)

Figure 2.5. The Giles self-propelled projectile with inventor Lloyd Giles. (Courtesy of National Air and Space Museum archives.)

enough with the four successful flights to keep an active interest in aerial torpedo research.

The U.S. Army Air Service continued to work on improving the aerial torpedo. The Sperry company designed a subscale one-seater biplane named the M-1 Messinger to test control mechanisms for other aerial bombs. Lawrence Sperry lost his life flying this aircraft when it crashed in the English Channel in 1923. The U.S. Army Air Service at Wright Field in Dayton, Ohio, continued research on radio-controlled versions of the M-1 until the Great Depression of the early 1930s, when cuts in the defense budget ended all aerial torpedo work in the United States.

The British renewed their interest in pilotless aircraft during the 1920s, viewing them as cost-effective and as a possible alternative to some manned systems (see Figure 2.6). The Royal Air Force examined several different types of pilotless aircraft, including a gyroscopically controlled aerial target with a range of 20 mi (32 km); an aerial torpedo that would be launched from an aircraft and then radio controlled for up to 10 mi (16 km); and a radio-controlled pilotless attack plane that could drop its own bombs. A decision was made to build a gyroscopically controlled vehicle that would incorporate both target and single-warhead–carrying capabilities. Several test aircraft designed to carry a 200-lb (90-kg) bomb at a speed of 103 mph (165 kph) crashed into the sea. A clockwork control device was substituted with a radio mechanism that actuated the controls in a preprogrammed sequence. This change led to drastic improvements, resulting in a successful test flight of the aircraft. A midwing monoplane production model incorporating a 200-hp Armstrong Siddeley Lynx engine that could propel the aircraft up to speeds of 193 mph (310 km/h) was chosen. This was much faster than the manned aircraft of the era. Twelve of these aircraft, called the LARYNX, for Long-Range Gun with Lynx Engine (a misnomer), were built at the Royal Aircraft Establishment at Farnborough. The LARYNX was successfully test launched from both British warships and installations. Six LARYNXs with live warheads were tested at the RAF Station in Shaibah, Iraq, a desert area considered to be quite safe for such demonstrations. Three of the aircraft turned out to be duds, and a fourth made a beautiful flight over the horizon into the desert, never to be seen again. The British lost interest in the LARYNX and by the 1930s primarily devoted their efforts to radio-controlled targets.

The requirement for a target aircraft was brought on because of a dispute between the Royal Navy and Royal Air Force as to whether battleships were vulnerable to attacks from aircraft. Following Billy Mitchell's test sinking of several warships by U.S. Army Air Services aircraft, the British government (in particular, the Royal Air Force) began to question capital ship survivability. The Royal Navy insisted that the Mitchell test was not realistic in that the ships were not maneuvering or shooting in self-defense.

Figure 2.6. Drawings of British 1920s concept of radio-controlled pilotless attack plane, which could drop its own bombs. (Courtesy of National Air and Space Museum archives.)

The Air Ministry decided to test the ship gunners by building a radio-controlled target aircraft that would simulate attacks on the fleet. Three standard Fairey III F reconnaissance floatplanes were reconfigured into radio-controlled vehicles named Fairey Queens. Much to the glee of the Royal Navy and to the embarrassment of the RAF, the first two Fairey Queens crashed into the ocean on their maiden test flights. In January 1933, the remaining radio-controlled floatplane was successfully launched and used as a target for the home fleet patrolling in the Mediterranean Sea. It survived more than two hours of concentrated naval gunfire before being safely recovered. The Royal Navy learned some very important lessons and realized that they had to make technical and operational changes in order to counter the new threat from the air. Following this incident, the British developed an all-wood target version of the DeHavilland Tiger Moth biplane, renamed the Queen Bee, of which 420 were built between 1934 and 1943.

World War II

In the mid-1930s, U.S. movie star Reginald Denny, who was also a model aircraft enthusiast, started the Radioplane Company in order to produce aerial targets for the U.S. military. Although radioplane's early prototypes were failures, by 1939 they had developed a very effective high-wing monoplane with two contrarotating propellers mounted side by side. This RPV, known as the RP-4, was much slower than the Queen Bee, flying at only 60 mph (96 kph) as compared to the 109 mph (175 kph) speed of her British cousin. Radioplane's aerial target, however, operated at a much lower altitude, which for gunnery practice offset its slower speed. The U.S. Army Air Corps initially purchased only a few, but by 1942 the United States was in the midst of a major war and both the army and navy ordered a large number of vehicles, totaling 984 of the OQ-2A version, 9,403 OQ-3s, and 3,548 OQ-13s. The OQ-3 and OQ-13 versions incorporated larger engines and could reach speeds up to 140 mph (225 kph) and altitudes up to 10,000 ft (3,050 m).

An air-launched, radio-controlled aerial bomb capable of carrying 500 lb (225 kg) of explosives for a distance of 50 mi (80 km) was built by General Motors at the onset of World War II. Fifteen of these bombs, known as the GM Bomb Bug, were procured; twelve of them were destroyed in tests. The U.S. Army Air Force (USAAF), determining that in order for this weapon system to be effective they would have to buy an excessive number of Bomb Bugs, decided that other projects showed more promise. Remotely piloted vehicles were operationally used by USAAF during World War II (Figures 2.7–2.9). A few flights were made in the Pacific theatre by remote-controlled piston-engined monoplanes carrying large bomb loads. In one bombing mission in Europe, B-17s dropped over 200 specially built 2,000-lb (900-kg) bombs that had plywood wings and fins attached to them. The Germans mistakenly believed that they had shot down most of the fighter escort, which exploded violently upon impact. They evidently confused the bombs for the fighters. USAAF also converted damaged B-17s and B-24s into remote-controlled bomb-laden aircraft. The pilot would fly the aircraft from takeoff until he approached the coast, at which point he would bail out while still over land. The plane would continue on under remote guidance to attack Nazi targets on the continent. It was this sort of aircraft, a B-24 loaded with TNT, that Joseph Kennedy was killed in when it exploded over England before he could bail out. This program was eventually cancelled, due to the high cost and technical difficulty of converting such large multiengined bombers to remote control.

In 1942 and 1943, USAAF conducted wind-tunnel tests at Wright Field on one-fourteenth-scale models of ground-launched controllable bomb airplanes. Two versions of twin-engined aircraft, the XBQ-1 and XBQ-2a,

Figure 2.7. World War II vintage experimental USAAF RPV, which was radio controlled by a pilot located on the ground wearing a breastplate with a steering wheel attached to it. (Courtesy of Dowty, RFL Industries.)

were specifically designed as remote-controlled bomb airplanes that would explode when they crashed into the target. They were planned to be television equipped and radio controlled, which would allow a remotely located person (probably on another airplane) to maneuver the bomb airplane into position. USAAF decided against producing specially built aircraft for the remote-control mission and opted to continue to convert manned aircraft.

No unmanned aircraft of World War II has received as much publicity as the Fiesler 103, better known as the V-1 Buzz Bomb. Although early prototypes of this weapon were available in the late 1930s, the V-1 was not given the go-ahead for production until 1941. The V-1 augmented the V-2 rocket as the cheap, simple weapon-system component of the "high–low mix." The buzz bomb was 27 ft (8.24 m) long and weighed 4,800 lb (2,160 kg), of which over two-fifths was devoted to its warhead. It flew at altitudes between 1,000 ft (305 m) and 7,000 ft (2,135 m) and at a speed of slightly over 400 mph (640 kph). It had a range up to 250 mi (400 km). The V-1's guidance system consisted of an Askania gyroscope used for direction and altitude. A small propeller, located on the nose, was driven by the bomb's passage through the air and served to actuate a primitive distance recorder. The fuel was automatically shut off at a preset distance, at which time the V-1 would initiate its dive, making the "buzz" sound its nickname is derived from.

Figure 2.8. Ground operator's control breastplate. (Courtesy of Dowty, RFL Industries.)

Ten-thousand, five-hundred sorties were launched against England from June 12, 1944, through March 30, 1945 (Figure 2.10). Only 2,500 survived both mechanical failures and the enemy's defenses to penetrate to their target. Nonetheless, they caused 14,665 casualties. The V-1 was quite often shot down by air or ground fire. They were occasionally tipped over by the wing of a fighter, causing them to veer out of control and then plunge into the English Channel. Although the V-1 was more vulnerable and less effective than the V-2, it was much cheaper to build, costing 1,500 marks as compared to 10,000 for the V-2. Each V-1 utilized only 280 man-hours of slave labor, contrasted to 13,000 for the V-2. Although each weapon system carried a similar payload, the V-2 was much more devastating, due to the nature of

Figure 2.9. A long cable connected the breastplate panel to the crystal-controlled single-frequency 35-W transmitter that served as the one-way communications medium, using a vertical anttenna as a radiator. (Courtesy of Dowty, RFL Industries.)

Figure 2.10. V-1 descending on London. (Courtesy of Air Force Museum archives, Dayton.)

the impact created by its ballistic flight. A manned version of the V-1, named the Reichenberg, was introduced at the end of the war. This vehicle was an experimental test aircraft designed to iron out some control problems, rather than becoming a suicide bomber, as originally thought by some Allied observers. A higher-powered version of the V-1 with an increased range was being worked on at the end of the war. Its development greatly influenced some of the early cruise missile research in the United States and Soviet Union following the war.

The U.S. Navy released a conceptual drawing to the press on October 20, 1945, depicting one of a trio of pilotless jet aircraft that were claimed to be near introduction. The aircraft, named the Glomb, Gorgan, and Gargoyle, were described by the navy as "robot craft" and as "Heralds of a supersonic age where only the mind of man can match the speed of the deadly creatures his genius has conceived." The Glomb, the largest of these weapons and designated the LBE-1, was envisioned as a television-controlled aircraft that could withstand 300-mph in a 4-G dive. None of the trio ever made it into the navy's inventory.[7]

1950s to the Present

In 1951, the Ryan Aeronautical Company provided the United States with the first jet-engined target drone, the Firebee. In the spirit of its founders, Sperry Corporation also began converting many manned aircraft, including F-80 Shooting Stars, into drones. During the Korean War, the U.S. Navy made drones out of surplus F6F Hellcats, which were then loaded with explosives and guided into heavily defended Communist targets. Manned aircraft such as the AD-4 Q Skyraiders were used as control planes to remotely maneuver the drones.

During the 1950s, both superpowers briefly relied on cruise missiles as important components to their strategic forces. The air force had the Snark air-breathing intercontinental missile, which carried a nuclear warhead and was controlled by an internal celestial guidance system. The Snark had a range of 6,325 mi (10,120 km) and a maximum speed of 650 mph (1,040 kph). The U.S. Navy had their own strategic cruise missile, the Regulus. It could be launched from submarines, aircraft carriers, cruisers, and naval land bases and could carry a nuclear warhead up to 500 mi (800 km) at a speed of mach 0.87 or mach 1.1 in terminal dive. A target drone version of the Regulus was also built. The Soviet T-4A prototype system consisted of a cruise missile that could be launched into near-space altitude by a conventional rocket, and then would glide towards its target. Guidance was obtained by signals emitted from a submarine off the target's coast, or through a radio beacon already in place. Cruise missiles, however, were soon deemphasized in favor of manned bomber systems and ballistic missile development. The Soviets had also developed a supersonic reconnaissance version of the T-4A in the mid-1950s known as the Yastreb (Eaglette) reconnaissance-ELINT RPV. It was not deployed until the mid-1960s.

The loss of manned U-2 reconnaissance aircraft over the Soviet Union and Cuba in the early 1960s, and the political embarrassment derived from the Soviet Union's propaganda exploitation of captured U-2 pilot Francis Gary Powers, led to a major U.S. effort to reconfigure the Ryan Firebee aerial target into a reconnaissance drone. The first "photo-recon" flight conducted by an RPV occurred in 1963. The Strategic Air Command's 100th Strategic Reconnaissance Wing was deployed to Kadena, Okinawa, in 1964. The Wing launched Teledyne-Ryan AQM-34s from DC-130 Hercules aircraft flying along the coast of Communist China. These AQM-34s penetrated China's airspace and obtained high-quality photographs of military facilities and troop movement. The drones were recovered on the surface of the South China Sea. In 1965, the Chinese held a news conference displaying a shot-down U.S. pilotless reconnaissance aircraft. This was the public's first view of an advanced RPV performing missions too dangerous or politically delicate to be undertaken by people. The air force continued to introduce drones that could fly at increasingly higher altitudes, thus avoiding Chinese defenses. The Ryan AQM-34N could reach 60,000 ft (18,300 m), which was higher than what China's SA-2 missiles could intercept. The air force also experimented with flying drones at very low altitudes over China. Improved relations between China and the United States in the early 1970s resulted in the suspension of RPV overflights.

In the 1960s, the U.S. Navy managed the short-lived Drone Anti-Submarine Helicopter (DASH) program. The prototype vehicle was very difficult to control on the deck of a ship, causing much concern among the crew that it would easily crash and harm people or expensive equipment. Although it was unsuccessful, the DASH provided some innovative tactical thinking and introduced the idea of rotor aircraft as capable RPV platforms.

The high attrition of air force and naval aircraft lost during bombing attacks in North Vietnam due to the sophisticated surface-to-air missile systems installed by the Soviets led to some real innovation in RPV technology and operational usage (see Figure 2.11). Over 2,500 photo-reconnaissance missions, code named "Buffalo Hunter," were flown over North Vietnam. The reconnaissance RPVs had an attrition rate of only 4% and they approached their photo runs at very low altitudes. The fine quality of their coverage was exhibited when the Department of Defense showed Congress post–B-52 damage assessment photos of the December 1972 raids taken by AQM-34L and M drones (Figure 2.12). Approximately 500 missions were carried out by AQM-34 Q and R versions, which performed such duties as electronic eavesdropping, jamming enemy radar frequencies, and the dropping of chaff in the corridors through which manned aircraft

Figure 2.11. Firebee being launched from the ground. (Courtesy of Air Force Museum archives, Dayton.)

flew. These RPVs were launched from DC-130s and conducted their missions at very high altitudes. Upon completing their mission, they were recovered by helicopters that would snatch the parachutes deployed by the RPVs.

The Yastreb reconnaissance ELINT RPV served as the primary Soviet RPV system through the mid-1970s. It had an operational ceiling of 90,000–100,000 ft (27,500–30,000 m) and its speed could exceed 2,188 mph (3,500 kph). The Yastreb, like the T-4A, took off from a mobile rocket launcher. It could follow a preprogrammed path or be radio controlled.

In the years following the Six Day War in the Middle East, the Egyptians, in violation of the terms of the cease-fire, began to install heavy air-defense belts on the Suez Canal. These included Soviet-built ZSU-23-4 quad-mounted 23-mm antiaircraft guns; SA-4 shoulder-fired, low-altitude anti-aircraft missiles; and the SA-6 medium-altitude missile. A young Israeli intelligence officer worked out a plan to neutralize the Egyptian defenses if a shooting war should ensue. His scheme involved using RPVs rather than manned aircraft in the initial assault on the air defenses. Prior to the 1973 war, the Israelis were able to discreetly obtain from the United States a quantity of drones, including the Ryan AQM-91A, Northrop Chukars, and support equipment. When war did break out, the Israeli Air Force (IAF) began to prepare a counterattack in which the RPVs would be used to draw the fire, reducing the flak and missiles faced by manned aircraft that would follow. Israeli senior military officials, however, countermanded this plan and instead sent in the manned aircraft (consisting of F-4 phantoms and A-4

Figure 2.12. Example of photography taken by RPVs in Vietnam: an antiaircraft site. (Courtesy of Air Force Museum Archives, Dayton.)

Skyhawks) to spearhead the attacks. The results were catastrophic for the Israelis. The Egyptians shot down more than thirty Israeli planes during the first day.

The Israelis eventually regrouped and reverted back to their original plan. They sent their Firebees and Chukars ahead of the next attack of manned aircraft. The RPVs drew a great deal of fire. One RPV survived an attack by thirty-two SAMs, and managed to return safely to the Israeli lines. The Israeli F-4s and A-4s caught the Egyptians while they were reloading; they were thus able to knock out the missile defenses and gain control of the air. The Israelis then began to really experiment with their RPVs. They were employed for the reconnaissance of missile sites and troop movements, and as electronic countermeasure (ECM) devices for decoying and jamming.

The BGM-34-A RPV was used by the IAF for the delivery of explosives to Egyptian missile sites and armored vehicles. This RPV located the target through a T.V. camera and lens situated in its nose. The target was then passed on to the person at a console safely behind the lines. By examining the T.V. picture, recording flight and location, the individual then selected the target and fired at the target an AGM 65 Maverick missile carried by the RPV. The RPV's camera image was relayed to the camera housed within the Maverick missile, which then automatically guided the missile to its target. In December 1973, Israeli officials acknowledged their use of RPVs to deliver weapons during the October War.

Following the 1973 Arab–Israeli War, many nations tried to incorporate the lessons learned concerning the use of RPVs. Countries such as Israel and the United States began programs to produce mini-RPVs, systems that would be more difficult to shoot down and that would gather intelligence more discreetly. Alvin Ellis, an American engineer who worked for Israel Aircraft Industries and Tadiran, convinced the Israeli military to try a remote-controlled model airplane capable of carrying a small T.V. camera. This proposal resulted in the development of two Israeli mini-RPVs, the Scout and the Mastiff.

While the Israelis were pursuing a very concerted mini-RPV effort, the United States began a slow, incremental mini-RPV program. In the late 1970s, however, another "robotic" aerial weapons platform was receiving a lot of attention in the United States. Cruise missile proponents claimed that these stand-off weapons now had the precision guidance technology that could maneuver them around Soviet air defense systems and radars, and then enable them to strike within 30 ft (9.15 m) of their targets. They also believed that the Soviets would have great difficulty defending against swarming cruise missiles that could be launched from ground, air, and sea platforms. President Jimmy Carter seized on this opportunity and greatly expedited the cruise missile program. He also used it as a justification to cancel production of the B-1A bomber. The United States, during the 1970s, developed and, during the 1980s, procured and deployed five different types of cruise missiles, including the USAF/Boeing AGM-86B air-launched cruise missile (ALCM) used for nuclear land attack and launched from B-52 and B-1 bombers; the USAF/General Dynamics BGM-109 G ground-launched cruise missile (GLCM) furnished to both the United States and allied nations with theater nuclear weapons; and three models of the Tomahawk sea-launched cruise missiles (SLCM), termed the BGM-109, which is produced by both General Dynamics and McDonnell Douglas for the U.S. Navy (Figure 2.13). The TLAM/N nuclear version of the Tomahawk is for attacking ground targets, the TLAM/C is for conventional attacks against land-based targets, and the TASM is a conventional armed antiship cruise missile for destroying enemy surface ships and submarines. The Tomahawk is housed on submarines, destroyers, and battleships.

During the summer of 1982, Israel invaded Lebanon to rid the area of Palestinian guerrilla strongholds. This resulted in pitched battles with Syrian forces. The Israelis realized that their key to dominating Lebanese airspace was to rid the area of Syrian-manned, Soviet-built SA-6 missiles. The SA-6 is mounted on a modified tank chassis and has a range of 22 mi (35.2 km). It has

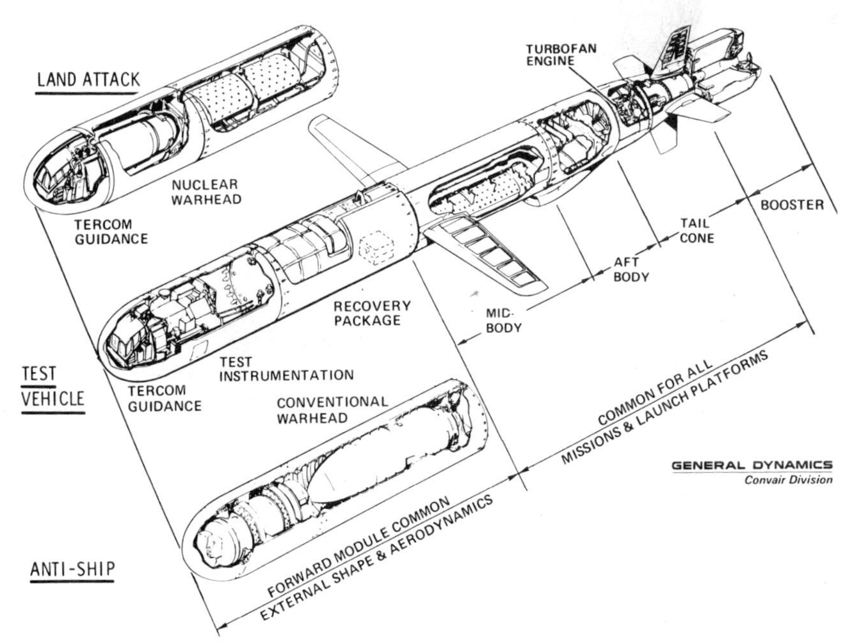

Figure 2.13. Sketch of the Tomahawk cruise missile and internal components. (Courtesy of General Dynamics.)

a sophisticated radar guidance and homing system, making it quite dangerous to aircraft, as the Israelis learned in October 1973. The Israeli-made Scouts and Mastiffs flew into the Bekaa Valley, emitting electronic images of Israeli fighter jets (Figure 2.14). The Syrian radars were activated, and the RPVs were able to pick up the radars' locations and signatures, and then transmit the information to nearby E-2 Hawkeye AWACS planes. The AWACS were then able to relay to Israeli aircraft the correct jamming frequencies. Chaff-dispensing rockets were used to obscure the radar images of the actual Israeli fighters and harass the Syrian air-defense crews. The Israelis were then able to launch from their attack aircraft radiation-homing or "smart" Wolf missiles, which devastated the SAM radar-guidance units. Being blinded, the Syrian SA-6 and SA-8 missile launchers and control vans became easy prey to the Israeli F-16s, which obliterated them with conventional and cluster munitions. The Israeli mini-RPVs also provided damage-assessment information and served to monitor the movement of Syrian armor and troops, and to keep various Palestinian guerrilla groups under close surveillance.

The notoriety that RPVs gained during the Bekaa Valley operations, as well as the attrition of manned aircraft experienced by the United States during operations in Lebanon and Grenada and by the British and Argentinians in the Falklands War, has spurred on a number of countries to emulate Israel's use of robotic aircraft.

During the summer of 1985, the U.S. Army revealed the crash of one of

Figure 2.14. Scout mini-RPV revealing camera underneath fuselage. (Courtesy of Israel Aircraft Industries.)

their Skyeye RPVs while on an intelligence-gathering mission in support of Salvadoran military operations. The U.S. Army had purchased Skyeyes built by the astronautics division of Lear Siegler as a stop-gap solution while awaiting completion of the more advanced Lockheed Aquila, which was experiencing numerous developmental problems. The U.S. Navy and Marine Corps, however, decided to procure the proven Israeli Scout and Mastiff systems, and their follow-up version, the Pioneer 1, to fill their mini-RPV roles.

MECHANICAL SWIMMERS

Early History

The European medieval intellectual renaissance produced some remarkable individuals who were well versed in the arts, philosophy, and science. Two such men were also creative visionaries who experimented with some rudimentary unmanned naval vehicles, and foresaw their future martial applications. Roger Bacon, the English philosopher and scientist of the thirteenth century, correctly predicted many upcoming technologies, including manned flight and the use of gunpowder. Bacon also wrote about a warship-like vessel that was experimented with in London. This machine is described as being able to travel wholly submerged through water and against waves at a speed greater than anything attained on land. Leonardo da Vinci claimed that he had also invented an underwater craft, but because he feared that the device was too destructive he decided against its further development.[8]

The idea of remote-controlled naval vessels does not reappear in historical accounts until 1885, when an electrically controlled boat was steered by an operator on board the British torpedo ship *Vernon*. The lack of suitable motors and the inability to transmit enough electric power through the cable made the full-scaled development of such a system impractical. At the turn of the century, both the British and the Germans experimented with remote-controlled torpedoes. The British Brennan and the German Siemen harbor torpedoes were conceived of as weapons that could be launched from land positions into the sea and used to defend harbors against enemy vessels. The required apparatus, however, became too bulky and complicated to be carried by torpedoes, and the research emphasis for remote control was shifted back to surface craft.

World War I

During World War I, the Germans felt confident enough in their remote-control technology that they actually built seventeen electrically controlled motorboats for coastal defense. These vessels, known as FL-7s, were 42 ft

(14.9 m) in length, had a beam of 6 ft (1.8 m), and were powered by 200-hp gasoline engines. They could reach speeds up to 30 knots, and carry enough fuel to travel for six hours. Their radius of action, however, was limited by their visibility to the operator sitting at the control station. These stations were placed on 100-ft (30.5-m)-high towers situated on the coastline, allowing the operators to see and direct each boat's maneuvers up to 15 mi (24 km) away from the shore. The boats were steered by electricity transmitted through an insulated single-core cable that could reach a length of 50 mi (80 km). The boats carried in the fore section up to 450 lb (202.5 kg) of explosives, which detonated upon contact, and, in theory, destroyed the suicide boat as well as the intended target.

By 1916, the control range of the German unmanned boats was doubled with the positioning of a seaplane that was equipped with a strong radio sender. The plane signaled to the shore operator the direction in which to maneuver the boat. The German destroyer T-146 was eventually configured to handle the boats by cable, thus eliminating the need for the shore stations.

Some German writers credit the remote-controlled boats with sinking a number of Allied vessels during World War I; however, no such claims have been substantiated by official reports. On March 1, 1917, an FL-7 did strike a British observation post on the shoreline and blasted a 150-ft (45.8-m) hole in the facility. Although the Germans considered the incident to have successfully demonstrated their new weapon, the British believed it to be foolhardy for the Germans to expose their secret weapon on a target of such limited worth. Remote-controlled boat attacks on Allied warships resulted in little damage, but they did help to keep British vessels away from the German coastline. On October 28, 1917, a motorboat directed by a plane flying overhead steered through a group of escorting destroyers and avoided being hit by the heavy artillery barrage laid down by British warships. The remote-controlled boat struck the heavy monitor *Erebus* amidships. Although the boat's explosives blew up on contact, the rail around the British ship reduced the impact, and only the bilge of the *Erebus* was damaged. The British monitor was repaired and back in service in less than two weeks.

Several factors reduced the overall effectiveness of the remote-controlled attack boats. These included the inexperience of team coordination between destroyer, seaplane, and land station; the fact that recurrent motor trouble reduced the number of operational boats; and that the Allies's net barricades often damaged the cables connected to the boats. Repeated failures resulted in a loss of confidence by the German navy for cable-controlled boats. This led to an emphasis on controlling boats through the use of radio signals emanating from the seaplane, which resulted in an extention of their radius of operations, up to 200 mi (320 km). Tactics were developed to use the boats in major offensive actions, and not just as a means of coastal defense.

Unfortunately for the Germans, World War I ended without an opportunity to fully test the possibilities of radio-controlled vessels. The very modest achievements obtained by the German navy did, however, demonstrate the feasibility of using remote-controlled craft in the battle environment.

U.S. Mechanical Swimmers

The remote-control boat technology employed by the Germans during World War I was experimented with and duplicated by the U.S. Navy in the early 1920s. In fact, the U.S. Navy placed a contract with Morkrum Kleinschmidt Company, the German developers of the teletype eight-circuit selector switches used in the remote-control World War I boat. Research incorporating this technology was conducted at the U.S. Naval Proving Grounds at Dahlgren, Virginia, but it was directed more to controlling remotely piloted aircraft than to maneuvering ships. After extensive testing, the teletype switches were deemed unsuitable for flight controls. The Naval Research Laboratory in Washington, DC, however, was able to apply the same technology to developing radio controls for target ships such as the U.S.S. *Stoddert* and *Utah*. During the next two decades, on through World War II, there was a surprising lack of research in remote-controlled naval vessels. This is even more startling when compared with the progress made on ground and air remote-controlled vehicles, which were developed and operationally used during the 1940s.

It was not until 1957 that the concept of using unmanned naval vessels was revived. A mobile underwater T.V. system built by the New Jersey-based Vare Company for commercial purposes caught the interest of the navy. Engineers at the U.S. Naval Ocean Systems Center (then the Naval Electronics Laboratory) developed the Cable-Controlled Underwater Recovery Vehicle (CURV) in 1958. The CURV recovered more than 600 torpedoes and other objects over the next five years, and generated a great deal of interest in unmanned submersibles, commonly referred to as Remote Operated Vehicles (ROVs). Use of such robotic systems was attractive to the navy because they could take some of the workload off of highly skilled personnel by performing dangerous missions that could entrap or disable manned vehicles or divers. From 1958 to 1974, the U.S. Navy funded development of eight more versions of the tethered type of ROVs, three of which were modifications to the CURV prototype.

Although the early unmanned submersibles functioned in noncombatant roles by performing such tasks as inspection, survey, search identification–location and retrieval, they did produce some dramatic results. For example, a U.S. Air Force KC-135 collided with a B-52 while attempting to refuel in 1966. The bomber was torn to bits, and one of the H-bombs it was carrying was lost in the Mediterranean Sea off the coast of Spain. The H-bomb's recovery was essential for reducing the political friction that arose between Spain and the United States, and to relieve the fears that U.S. nuclear weapons and technology could fall into the wrong hands. The bomb was located in 2,850 ft (869 m) of water by the manned submersible ALVIN. Concern that the ALVIN could become entangled in the shroud lines of the bomb's parachute if an attempt was made to retrieve it convinced the navy to try out their CURV I. Two lift lines were attached to the bomb by the ROV, but the CURV I became entangled while in the process of connecting the third lift line. Fortunately, the two lines were enough to pull both the bomb and the CURV I to the surface.

In June 1973, the manned Johnson Sea-Link (JSL) minisub became snared in a scuttled destroyer's rigging off the coast of Florida. Due to the strong currents that existed at the 360-ft (110-m) level of water, divers were unable to free the minisub. Another manned submersible malfunctioned while attempting a rescue and had to be recalled. After 32 hours, a U.S. Naval Ordnance-built ROV attached a line to the JSL, and it was successfully pulled free from the sunken destroyer. Just two months later, another manned submersible, the British PISCES III, sank off the coast of Cork, Ireland, to a depth of 1,575 ft (480 m). The U.S. Military Airlift Command flew a CURV III to Ireland to participate in the rescue (Figure 2.15). Other manned submersibles managed to attach two lift lines to the PISCES III; however, a third essential line affixed by the CURV III enabled the submariners to be rescued.

The U.S. Navy's CURV III aided the U.S. Environmental Protection Agency's Office of Radiation Programs in assessing the status of an underwater burial area used to dump radioactive wastes and chemical munitions from 1946 to 1959. The possibility of heavy contamination discouraged the navy from using divers or manned systems. Over a five-day period the CURV III was able to take hundreds of color photos of the area and grabbed sediment cores and container samples. Another U.S. Navy cable-controlled ROV named DEEP DRONE assisted in the salvaging of a 41-ft (13-m) Coast Guard boat off the coast of Oregon. In very severe weather the DEEP DRONE managed to attach a 7-in. (18-cm) hawser to the craft, which was resting in 315 ft (96 m) of water. Surface conditions were so rough that 15-ft (4.5-m) outriggers had to be attached to the support ship to keep the DEEP DRONE from colliding with it when being lowered or lifted from the water. In March 1977, a U.S. Navy A-7 aircraft crashed off the coast of Japan. The pilot was unable to eject and drowned, still strapped in his seat. The ejection seat and aviator were located in 340 ft (104 m) of water by the DEEP DRONE. The ROV was able to connect a grapnel hook to the ejection seat, which allowed a support ship to pull both ejection seat and the body to the surface.

The CURV III proved instrumental in overhauling the U.S. Navy's Fixed

Figure 2.15. CURV III hanging from a crane alongside a boat during testing. (Courtesy of U.S. Navy.)

Acoustic Range (AFAR). It rigged an acoustic tower in such a way as to enable it to be lifted from the sea floor. The CURV III also cut electric cables, retrieved them, and mapped and inspected with sonar the existing tower sites. In numerous other instances, tethered ROVs have inspected aircraft underwater crash sites and sunken ships, and have retrieved torpedoes and bombs.

The U.S. Navy's TORTUGA program involved the building of a much smaller vehicle than the CURV. It could be deployed from a submarine and used to maneuver an underwater video system in order to closely examine areas that up to then had been inaccessible. The first such vehicles used water jets for propulsion, but later switched to propellers for increased maneuverability. The ANTHRO program (standing for anthropomorphic) was a navy endeavor that somewhat resembled the U.S. Army Tank–Automotive Command and General Electric's walking machine program. Both developed robotic vehicles designed to duplicate the operator's movements and each may have influenced the other, as they took place around the same time period, in the 1960s.

The ANTHRO involved a technique referred to as "head coupled," in which the camera located on the unmanned submersible would orient itself through cabled instructions to the same relative position assumed by the operator's head, the operator being located on the support ship. The scene provided by the submersible's camera moved in synchronization with the operator's head movement. Research indicated that the operator could remember the relative location of most of the objects that passed by his field of view. Audio inputs obtained from hydrophones attached to the vehicle were relayed to the operator's ears in order to examine the feasibility of detecting and locating underwater objects. The operator wore a helmet that contained a 5-in. (12.9-cm) T.V. screen, roll, pitch, and azimuth sensors, and dual headphones. The operator also controlled the ROV's maneuvering, depth functions, and television focus via a control unit located at his right hand.

The SCAT (Submersible Cable-Activated Teleoperator) program was a follow-up to ANTHRO. It also served to examine head-coupled and three-dimensional television display technology. The two SCAT submersibles, nicknamed SNOOPYs, were small and lightweight and were meant to replace divers in some observation and surveillance tasks.

Other Research on Mechanical Swimmers

The Soviet Union also experimented with an unmanned submersible whose actions patterned those of the operator. The Soviet Institute of Oceanography built their MANTA vehicle in 1971. The MANTA was equipped with tenso-sensors that provided feedback to a special servo-controlled, hydraulically driven operator's chair. The chair was designed to repeat all the pitch and roll movements performed by the MANTA submersible, thus allowing the operator to experience the ROV's maneuvering. As of 1973, the MANTA system had been tested only under laboratory conditions. Little is known to the public as to where the ANTHRO and MANTA research efforts have led, but it is likely that they experienced the same problems that plagued the U.S. Army/GE walking vehicle program. Robotic systems requiring a great deal of the operator's physical interaction will lead to excessive amounts of stress and fatigue on the individual.

The University of Washington's Applied Physics Laboratory (APL), under navy funding, in 1963 began the development of a nontethered, free-swimming vehicle whose movements were unhindered by a cable. The electromechanical umbilical cord, or cable, used by most operational ROVs has the advantage of providing an unlimited energy supply, a high data transmission-rate capacity, and a safety link to the support ship. The cable, however, restricts the ROV's maneuverability and accessibility and often requires extensive maintenance and replacement. The APL program resulted in the development of the SPUR (Self-Propelled Underwater Research Vehicle), which was an untethered, controllable-trajectory ROV that gathered data on physical properties of the sea, such as temperature and sound velocity, and investigated submarine wake phenomenon (a distinctive water motion produced by submarines). The support ship communicated to the SPUR through a transducer array that sent frequency-shifted and digitally coded signals through an acoustic link.

In the mid-1970s cable-controlled ROVs were introduced to the tactical arena as mine-destruction devices. One of the first such vehicles was the PAP 104 built by Société ECA of France. Over 250 units of this system have been built and sold to ten different navies. The British used the PAP 104 during the war in the Falklands. It effectively carried out thousands of mission runs. The PAP 104 destroyed mines by either carrying explosive charges to detonate them or by using cutters to slice their anchoring lines, thus floating them to the surface for detonation. The Royal Navy used a Sea Owl ROV built by the Swedish firm SUTEC during the summer of 1984 mine-clearing operation in the Red Sea. This vehicle succeeded in locating and recovering a Soviet mine never seen before. Following the successful and cost-effective use of ROVs to clear mines, a number of countries, including the United States, began to develop similar or enhanced ROV capabilities.

OUT-OF-THIS-WORLD ROBOTS

As we approach the development of the space station, orbiting laser-battle stations, and transatmospheric vehicles (TAVs or "space planes") there are some visionaries who believe that we are entering an historic period in mankind's evolution. They view these developments as a prelude to the establishment of manned lunar bases, planetary colonies, and eventually, humanity's migration to the stars. Other futurists, however, view man as being too biologically frail, with lifetimes too short to endure the rigors of space travel and settlement beyond "Mother Earth." They propose using robots to serve as our surrogates into the heavens. Arthur C. Clarke has written, "Creatures of flesh and blood like ourselves can only explore space and win control over an infinitesimal fraction of it. Only creatures of metal and plastic can ever really conquer it."[9]

The earliest satellites and space probes used innovative remote control and preprogrammable technologies in order to accomplish their missions. Until recently, space had been relatively free of weapons. This excludes strategic missiles (ICBMs and SLBMs), which encounter space only as a short-term medium (a matter of minutes) while accomplishing their ballistic arc and subsequent descent of warheads. Since the early 1960s, the superpowers have developed robot satellites to eavesdrop on each other's secrets through sophisticated photography and electronic surveillance systems. In addition, satellites have transmitted sensitive military communications around the world.

Beginning in 1967, the Soviets began testing devices for the targeting, interception, and destruction of satellites. The last such test took place in June 1982, when the Soviet hunter-killer satellite *Cosmos* 1379 was launched from the Tyuratam space launch center. On its second orbit of the Earth, *Cosmos* 1379 intercepted its prey, *Cosmos* 1375. Instead of signalling a destruct signal to blow up both *Cosmos* satellites, the Soviets commanded *Cosmos* 1379 back to earth to burn up in the atmosphere. On several earlier tests the Soviets did, however, destroy the hunter-killer satellites with conventional explosives to test their ability to destroy target satellites.

The first true space robotic system, although not a military system, was the U.S. *Surveyor* series of unmanned lunar landers. It had a pantograph arm to scrape up soil samples up to 3 ft (92 cm) away. The arm, capable of rotating in a 112-degree arc, could reach from up to 40 in. (102 cm) above the lunar surface to 18 in. (46 cm) below. In the first instance of celestial repair by a robot, the *Surveyor* 7 was able to use its arm to dislodge a stuck instrument. The Soviet lunar landers used a less-sophisticated boom to drill for core samples. The Soviet Union, however, was the first nation to remotely control a ground robotic vehicle on a nonterrestrial surface. Soviet operators based on Earth were able to remotely control the *Lunokhad* rovers, by viewing the cratered surface through stereo cameras mounted on these robots.

By the time the United States had reached the planet Mars with its two *Viking* landers in 1976, scientists had really perfected the development of a sophisticated robot arm that was actually rolled up during storage, but when flexed, formed a rigid structure. This arm scooped up Martian soil up to 8 ft (2.4 m) away, and then took the samples to an opening on top of the spacecraft that led to an automated analysis lab.

The space shuttle flights, which began in 1981, have been equipped with an advanced robotic arm named the remote manipulator system (RMS). The Canadian government developed the arm at no cost to the United States, in return for NASA's purchase of one RMS for each shuttle from the Canadian prime contractor, SPAR Aerospace. SPAR developed the arm and control panel, while Rockwell International built the RMS positioning mechanism and retention latches, and IBM supplied the control software. This 50-ft. (15.2-m) arm deploys and retrieves up to 65,000 lb (29,485 kg) from the cargo bay. It is operable in both automatic and manual modes and has been used to support EVA ("space walk") activities and payload servicing. The RMS arm houses lights and T.V. cameras to aid the mission specialist in its maneuvering. Even in light of these accomplishments, a 1980 NASA study group of machine intelligence and robotics chaired by noted astronomer and author Carl Sagan concluded that NASA's programs were five–fifteen years behind the state of the art in computer science technology and robotics.

Following the tragic loss of the U.S. space shuttle *Challenger* with its crew of six astronauts and schoolteacher Christa McAuliffe on January 28, 1986, there were renewed arguments for using more unmanned spacecraft. This event may influence whether the next generation of space craft, being designed to take off from a runway and ascend into orbit and then return to Earth like a conventional aircraft (the TAV), will be manned or robotic.

For very different reasons, the same sort of debate is emerging in the development of new ground-based weapon systems. Important decisions will be made in the future that determine whether man or machine will predominate in ground-based combat. In many respects, these debates will be necessary because of the increasing lethality of the future battlefield.

NOTES

1. Asimov, I. and Frenkel, K. Robots. *Machines in Man's Image.* Harmony Books, New York, 1985, p. 384. Accounts of both Pope Sylvester's and Rabbi Loew's flirtations with artificial men are described in this book.
2. This has become a reccurring theme in science fiction, in plots exhibited by the late-1970s T.V. series, *Battlestar Galactica*, and the 1984 movie, *The Terminator*.
3. Asimov and Frenkel, *Machines in Man's Image*; and Asimov, I. "The Perfect Machine." *Science Journal*, October 1968, p. 116. Both references discuss Asimov's Three Laws of Robotics, which are no longer just depicted in science fiction novels but form the basis for serious discussions in technological policy texts.

4. Clark, R. *The Role of the Bomber.* Thomas Y. Cromwell Company, New York, 1977, p. 9. Clark mentions the Hindu poem and its depiction of the manless winged chariot.
5. Clark, *The Role of the Bomber,* pp. 9 and 10; and Finnegan, J. *Military Intelligence: A Picture History.* U.S. Army Intelligence and Security Command, Arlington, VA., 1985, p. 10. These references furnish more recent historical records on unmanned aerial vehicles including military balloons and kites.
6. Smith, H. F. ("Red"). "From the Kettering Bug . . . The World's First Cruise Missiles." *Remotely Piloted Magazine,* January 1978, pp. 3–4; and *Air Force* magazine, October 1977. The U.S. Navy described aerial torpedoes as "automatically controlled aeroplanes or aerial machines carrying high explosives capable of being initially directed and thereafter automatically managed."
7. A photo of the Glomb accompanied by the U.S. Navy press release is on file in the Smithsonian Air and Space Museum's photograph archives, in the section on RPVs. Anyone experienced at reading contemporary DoD press releases would find the language to be quite unusual and refreshing.
8. Sweeney, James B. *A Pictorial History of Oceanographic Submersibles,* Crown Publishers, New York, 1970. Sweeney describes both Bacon and da Vinci's experiences with underwater craft, as well as some of the early manned diving and submerging equipment.
9. Gallagher, Edward J. *A Thousand Thoughts on Technology and Human Values.* Humanities Perspective on Technology Program, Lehigh University, New York, 1970, p. 47. This book contains Clarke's statement on the robotic exploration of space.

3 Current Operational Use and Development of Unmanned Robotic Ground Vehicles

If unconstrained, today's conflicts may become, and in the near future will certainly be, far too deadly for human beings to survive on the battlefield. The lethality of modern weapons, including nuclear–biological–chemical (NBC) munitions, hypervelocity missiles, smart bombs, lasers, and other high-technology killing mechanisms, are rapidly eroding what an individual soldier's initiative or heroics can contribute towards winning or losing. Whether soldiers die or survive has as much to do with political considerations that determine the level of conflict escalated to as with the tactics and valor employed. Clearly, much of the future battlefield will be too hazardous for people to operate in.

The U.S. Army has realized that there is a need to replace soldiers with robots in certain high-risk combat environments. Good soldiers are too few and too vulnerable to be used on missions with little survivability. Our values and beliefs prevent us from willfully sacrificing soldiers, even those who are inefficient or lacking in combat ability, in suicide missions. Yet to win battles, many of these missions must still be undertaken. A solution to this dilemma is to carry out many of these functions with robotic systems. The "think tank" SRI International developed for a U.S. Army report, "Army Applications of Artificial Intelligence/Robotics," a list of 100 different missions in land warfare for which it recommended the use of robots to replace or complement soldiers and manned systems.[1]

Although there is now a general consensus among most military strategists that unmanned systems will play an increasingly important role on the battlefield, this view was not easy to arrive at. For years the military bureaucracies, headed up by officers whose careers were advanced by their proficiency in operating vehicles or by their leadership over men, resisted the use of systems that countered the wisdom derived from their past experiences in personnel-intensive armies. It also threatened the institution's power base by taking financial resources and leadership billets away from manned systems. This reluctance might have been for the best, because it allowed extra time for unmanned systems technology to mature. The lethality of the weapons used in recent wars in the Middle East and the Falklands has led the military leadership to support a battlefield role for robot vehicles. In the Syrian–Israeli tank battles, scores of armored vehicles engaged in combat as dense and vicious as anything seen in World War II, yet many more were decimated much more quickly due to the introduction of precision targeting and antitank weaponry than were the tanks of thirty years earlier. The effectiveness of aerial remotely piloted vehicles has also served to spur on the acceptance of land robotic vehicles.

The efficacy of using robotic vehicles is no longer in question. The debate now focuses on the question as to what extent men should be "in the loop" in their command and control. Technologists are developing artificial intelligence hardware, software architecture, and sensor technologies to allow for the fielding of autonomous vehicles by the turn of the century. Much of the military leadership, however, is aghast at the idea of vehicles that operate independently of their command and control and that may run amok. To appease these concerns, many of the weapons laboratories are emphasizing the development of teleoperators that maintain the man in the loop. These vehicles are remotely controlled by operators who rely on sensory input to guide their movements. The advantages and drawbacks of remote-controlled, teleoperated, and autonomous robots will be expounded upon later in this chapter.

U.S. MILITARY PROGRAMS — OVERVIEW

The U.S. Army is examining robots in a number of programs that range from the simple remote-control variety to extremely sophisticated and technically complex autonomous systems. Their Tactical Reconnaissance Vehicle (TRV) program is an effort undertaken by a team of organizational members, including the Human Engineering Laboratory, the Tank–Automotive Command, the Engineering Topographic Laboratory, the Night-Vision Laboratory, and the U.S. Army Armor Center to develop a whole family of combat robotic vehicles. The TRV involves putting different modules on an existing army chassis to perform a variety of missions, including tactical reconnaissance, weapons firing, smoke laying, and nuclear, biological, and chemical (NBC) detection. This program allows the army to study issues of force effectiveness, mobility, and communications. The TRV would start off largely as a teleoperated vehicle that would evolve into more of an autonomous vehicle by incorporating technologies developed from DARPA's Autonomous Land Vehicle program. These technologies include on-board route planning, terrain analyzing, driving aids, and remote-control software.

The U.S. Army's Armor Center at Fort Knox has drafted an operational and organizational plan recommending that a family of robotic combat vehicles be developed from a generic chassis that could carry a variety of mission modules, depending on the specific mission. These modules could provide for direct-fire capabilities, intelligence gathering, and obstacle breaching and emplacement. The near-term plans involve development of vehicles that have a certain amount of autonomous operation, but require a human operator to teleoperate the vehicle into critical positions and maneuvers. This concept envisions a number of teleoperated vehicles with some preprogrammed instructions incorporated into their on-board computers, which are controlled by human operators located in a command and control vehicle.

The U.S. Army's Military Police School has initiated a project called the Robotic Observation Security Sensor System (ROS^3). The goal of this program is to develop a sophisticated intrusion detection and response system that relies primarily on mobile robotic vehicles equipped with state-of-the-art sensors. Cost is the main factor for developing such systems. The school estimates that when a robotic vehicle replaces a manned guard post, it will result in savings of at least 50% calculated over the life of the vehicle. A robot of this type would cost between 150,000 and 200,000 U.S. dollars. The cost of manning a guard station is between $80,000 and $125,000 per year. The robot's life is expected to be about twelve years.

The U.S. Army Infantry School and the Missile Command are developing a Robotic Anti-Armor System (RAS) that has a remote missile system mounted onto a mobile chassis. The operator can be located safely away from the robotic vehicle. Its use would eliminate the current feeling that missile operators have of being threatened and extremely vulnerable during firing, which harms their concentration, frequently resulting in a missed shot. The missile operator can stay locked onto the target without exposing himself to return fire. It would greatly improve the effectiveness of antitank missiles. The RAS could also be used in areas that may be tactically advantageous, but with no clear avenue of retreat. Suicide missions would be acceptable for a disposable robotic system that could act as a force multiplier by taking out a number of tanks before being destroyed.

The Defense Advanced Research Projects Agency, in conjunction with the U.S. Army Artillery School, has been working on Future Artillery Systems Technology (FAST). Although part of the program is looking at autoloaders, another facet is examining the "remoting" of artillery platforms. By using autonomous ammunition handlers to transfer shells in conjunction with artillery platforms that are remotely operated, one could increase the rate of fire, plus eliminate the loss of personnel from enemy counterfire. Combining this system with fire-finding radar (which tracks incoming shells) and then extrapolating the firing point can produce a very devastating robotic weapon system.

The U.S. Marine Corps has established a Program Management Office for Ground–Air–Telerobotics Systems (GATERS) at the Development Center in Quantico, Virginia. The mission of the office is to define technical requirements and provide Marine Corps management for emerging systems.

MAN-IN-THE-LOOP KILLER ROBOTS

Remote-controlled vehicles are those in which the operator relies on either his own line-of-sight vision from a distance, or sensor data to maneuver a vehicle to perform a certain mission. *Teleoperators* are the most sophisticated type of remote-controlled vehicles that rely on sophisticated sensor systems. Teleoperators enable the military chain of command to have continuous control over the vehicle's movements. The machine has no intellect of its own and it simply mimics the actions of the operator. A number of disadvantages are associated with the use of teleoperators.

1. If the communications between the operator and vehicle are jammed or disrupted or the cables are cut, then the vehicle loses its functional utility. Modern battlefield conditions will pose a severe threat to the link between the man and the robot.
2. Operator-controlled machines may perform certain activities much more slowly than a robot relying on machine intelligence.
3. The operators may become highly prized targets and if they go, so may a number of teleoperators.

With these limitations in mind, there are certain missions in which a teleoperator would be most suitable.

One such mission is covering-force area operations, which involves being immediately in front of the main concentration of troops. Teleoperators would be useful in providing some reconnaissance intelligence to keep the field commander abreast of developing military situations. The covering force, consisting of remoted missile systems, would be able to help bear the brunt of the enemy's armor attack, thus furnishing time for the main forces to prepare for the counterattack.

Another mission in which teleoperators could prove useful is in retrograde operations, in which the outnumbered defending force tries to slow down and cause attrition among the advancing force. Teleoperators would be expendable and yet provide the needed time to allow for the remainder of the force to safely pull out. Teleoperators would not be useful in missions that require a great deal of penetration into enemy lines or in operations consisting of intermingling among dense enemy forces, due to the weakness of the command control link with the human operators. The only way to accomplish these functions is to have operators accompany the teleoperators in command vehicles. Otherwise, such offensive operations may be better left to autonomous systems.

The Fire Ant

The Standard Manufacturing Corporation has developed a low-cost, radio-controlled 1,500-lb (680-kg) explosive-filled vehicle called the Fire Ant (Figure 3.1), which can move at speeds up to 85 mph (138 kph). An operator located in a remote area directs the Fire Ant into a kamikaze-like collision with an enemy tank. The Fire Ant's low profile of only 3 ft (0.925 m) in height makes it a very difficult target to hit. Standard Manufacturing Company also manufactures the 4,300-lb (1,948-kg) Remote Control/All Terrain Vehicle (RC/ATV) which is controlled by an operator up to 3 mi (4.8 km) away. It can be configured with a variety of weapon systems and detection equipment.

Figure 3.1. Fire Ant with its six off-road racing motorcycle tires. (Courtesy of Standard Manufacturing Company.)

The Sprinkler

Universal Military Robot Corporation, founded in Longmont, California, in 1984, is designing an unmanned mobile weapon called the Sprinkler. Its name is derived from its ability to indiscriminately spray the surrounding area with machine-gun fire at a rate of 720 rounds per minute. This 4-ft, 2-in. (1.3-m) device is very inexpensive (around $2,000) for a remotely controlled robot; because it is not selective in what it is shooting at, it thus does not need an electronic brain. It would be useful only in situations in which commanders are certain that they want to kill everyone in its immediate vicinity.

The Prowler

The Programmable Robot with Logical Enemy Response (PROWLER) (Figure 3.2) is a multipurpose land robot produced by Robot Defense Systems of Thornton, Colorado. The basic vehicle is remotely controlled with some evolving semiautonomous capability. Designed for a variety of uses, the PROWLER is a 4,000-lb (1,816-kg) all-terrain vehicle powered by a diesel engine. This six-wheeled vehicle can carry a 2,000-lb (907-kg)

Figure 3.2. The six-wheeled, all-terrain PROWLER, outfitted with twin M-60 machine guns and grenade launcher. (Courtesy of Robot Defense Systems.)

payload at a maximum speed of 17 mph (27 kph). Its low-pressure tires provide the advantage found with wheeled vehicles, including less required maintenance, greater speed, and less expense than tracked vehicles; yet it also furnishes some of the benefits associated with tracked vehicles, including low pressure and excellent traction without the pavement damage caused by tracks. All six wheels are driven by way of a hydraulic lock and skid system. This design allows the vehicle to operate even when there is a damaged tire on both sides.

Vision

The PROWLER has on board a Motorola 68000-class computer with 32-bit microprocessors, which can control a large variety of optional sensors. A central operator in a remote location can maintain visual contact through three video cameras. One of the cameras is part of a complete vehicle diagnostic monitor system. Using a split-screen monitor, the vehicle's speed, fuel and oil levels, temperature, and pressures can be continuously observed. The same monitor can display the location, direction, and ammunition supply statuses. The second camera, mounted in the turret, can rotate 360 degrees. Optional weapons are mounted coaxially with this camera. In this manner, the camera can be used for visual navigation and fire control. The third camera, located on top of the turret, can also rotate 360 degrees, but in addition can be extended 30 ft (9.1 m) high on its telescoping mast. This function, combined with 30 degrees of vertical movement, allows for observation over the top of obstacles such as hills, trees, vehicles, and other structures. Night-vision options complete the visual capabilities of the PROWLER. Audio systems include directional and nondirectional microphones and a speaker for two-way communication between the operator and subject when within range of the vehicle.

Navigation

Navigational sensors for the PROWLER include laser range finders, directional gyros, and distance-measuring equipment. Multiaxis vehicle attitude sensors relate the steepness of terrain. On the battlefield, armor-impact sensors can detect projectiles hitting the vehicle. Infrared scanners and doppler radar, with additional assistance from an electromagnetic motion detector, actively look for opposing forces. A stationary PROWLER can detect moving targets in foliage and can operate a seismic monitor that detects subaudible man- or vehicle-made ground vibrations.

Weapons

A wide range of armaments have been mounted and tested on the PROWLER. Options depend on the specific mission and can include both lethal and nonlethal weaponry. Nonlethal weapons consist of semiautomatic shotguns that fire a variety of low-lethality ammunition, including rubber slugs and multiball rubber projectiles. A multiround grenade launcher capable of firing twelve 40-mm tear gas bombs in two seconds is also available. A sonic disrupter that would emit an extremely high decibel sound in order to induce auditory pain, fear, and physiological disruptions is currently in the experimental stage. The PROWLER's diverse lethal weapons range from small arms to guided missiles. Nine-mm submachine guns can be tandem mounted in the turret. A range of light machine guns, from the Maremont M-60, which fires 650 rounds a minute, to the General Electric M-134 Minigun, which fires 6,000 rounds a minute, are other possibilities. Single-mounted weapons include the Maremont 50-caliber heavy machine gun, the Hughes 30-mm cannon, the Waterveliet Arsenal 105-mm recoilless rifle, and the Mark 19 40-mm grenade launcher with a firing rate of 350 rounds a minute. Missiles that can be mounted on the PROWLER include the TOW I and II systems, Hellfire tactical missile, Viper light antitank missile, and Stinger guided missile. The M9E 1-7 flamethrower completes the arsenal.

Uses

As a remote-control robot, the PROWLER uses a bidirectional radio control from a central human operator. The operator can be up to 19 mi (30 km) distant; the PROWLER's cruising range is 155 mi (250 km). As a sentry robot, the PROWLER can guard a perimeter as a stationary observer or through continuous movement. It can also be programmed for a combination of movements interlaced with periods of stationary duty. The sensors are programmable as well. The PROWLER can detect an intrusion and then move closer for further identification and confirmation. It can respond in a variety of ways, from synthesized-voice warnings to armament use. If desired, identifications can be further verified by a human operator. The need for manpower can be greatly reduced with a sentry robot like the PROWLER, which can tirelessly patrol large areas around the clock. See Box 3.1 for a PROWLER scenario.

Both DARPA and the U.S. Army's Ninth Infantry Division have awarded contracts to RDS to furnish field demonstrations of the PROWLER's capabilities. The Defense Electronics Division of Gould, Inc. teamed up with RDS to bid on a U.S. Army proposal for robotized tanks that incorporate the PROWLER's technology. RDS has furnished analysis to both Bell Aerospace Division of Textron Inc. and Boeing Aerospace Company on the use of PROWLER-type robots for the security of intercontinental ballistic missile installations. The PROWLER has also been proposed for security work at an installation in a Middle Eastern country by Bechtel National, an international construction firm.

PROWLER SCENARIO

Box 3.1. While on a preprogrammed patrol pattern of a nuclear missile silo, the robot's electromagnetic motion detectors sense someone scaling the fence. The PROWLER informs the operator of an intrusion, at which time the operator takes remote control of the vehicle, directing it with a joystick. The operator moves the PROWLER closer to the intruder to view the area with its camera system. Once the perpetrator is visible, the PROWLER allows the operator to talk to the intruder through its audio feedback with directional pickups. If the intruder is uncooperative and appears to have terrorist objectives in mind, the PROWLER can then use either nonlethal or lethal weaponry to disable him. The PROWLER has the added feature of providing a continuous video-recording capability to document the incident.

The TOV

The U.S. Marine Corps plans to field its Teleoperated Vehicle (TOV) by 1990 (Figure 3.3). The program is managed at Quantico, Virginia, by the Program Management Office for Ground–Air Telerobotics Systems. Its design philosophy is centered around building a vehicle that can be easily controlled by a minimally trained operator. The operator, situated as far as 18 mi (30 km) from the robot, wears a helmet that mounts stereo-optic video receivers and headphones in order to allow him or her to see and hear some of what a driver of the vehicle would actually experience via connected fiber optic cables. The operator's mock-up contains controls similar to those on the vehicle. A two-camera stereo-optic system is located on the TOV and furnishes the operator with a three-dimensional image and depth perception. The TOV's pair of binaural microphones allows the operator to hear sound in the direction it is received at the vehicle. The TOV's cameras and microphones are manipulated by the head movements of the operator; thus the sensors provide input from what the operator wants to focus on. This system is very easy to operate. Anyone who can drive a car can also put on the helmet and remotely operate the TOV with as little as ten minutes of practice. The disadvantage of this type of system is that the robot's performance is dependent on the alertness of the operator, and the operator may quickly tire from receiving all this "instrumented" input. The use of fiber optic cable is also questionable. Although it allows for secure communications free from interception and for a high rate of data transmission, the line may become cut by other vehicles crossing over the fiber optic's path or by the explosives or shrapnel that would exist in a combat environment.

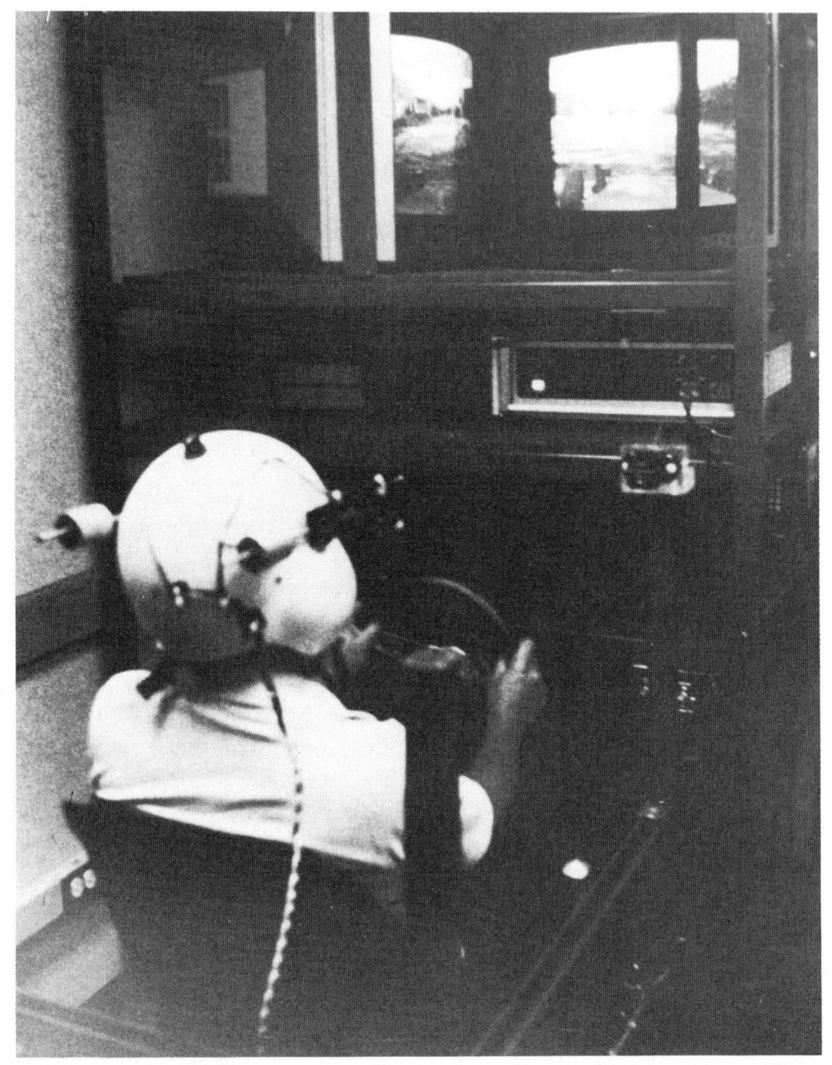

Figure 3.3. Marine Corps operator of teleoperated vehicle shown relying on visual and other sensory feedback to control the unmanned system's maneuvering. (Courtesy of of U.S.M.C.)

These weaknesses will be examined during initial field testing to determine what real problems they pose.

The vehicle initially being used for the TOV platform is the U.S. Army's Fast-Attack Vehicle (Figure 3.4), which is a "militarized" dune buggy. The GATERS (Ground–Air Telerobotic Systems) management, however, plans to adapt the TOV to the larger High-Mobility Multi-Purpose Wheeled

Figure 3.4. A fast-attack vehicle altered into the TOV test vehicle. Vehicle houses various sensors to furnish information to the operator located some distance away. (Courtesy of U.S.M.C.)

Vehicle (HMMWV) prior to being fielded. The vehicle will have an on-board navigation system that will relate the vehicle's position in military grid coordinates to the operator.

Three different missions are planned for the TOV, each with its own module.

1. In the reconnaissance mode, the TOV will have an elevatable mast that allows sensors to reach a sufficient height for long-range surveillance in the defilade position. Sensors include a day/night T.V. camera and a laser designator and range finder.
2. In the area-suppression mode, the TOV will be outfitted with a universal weapons mount that could carry the MK 19 grenade launcher, M-2 machine gun, or M-60 machine gun.
3. In the antiarmor mode, the TOV carries a TOW I missile launcher and associated sighting and tracking systems.

Battlefield Use of Teleoperated Robots

There have been several interesting proposals for using teleoperated robots on the battlefield. An approach conceived by the British robotics pioneer M. W. Thring has coordinated action among a regiment of teleoperated robot soldiers controlled by a manned tank equipped with the necessary command and control gear. The manned tank would be located several miles away in a well-protected spot. In a jungle scenario, Thring envisioned that a group of these robots would be spaced in a line 65.6 ft (20 m) apart and move at about 9.3 mph (15 kph) through a dense foliage, destroying all enemy resistance. Other robots, outfitted with horizontal chain saws, could level trees and plants to clear paths through the jungle.

Robert Finkelstein, president of Robotic Technology Inc., has proposed a concept he calls Troika, which is a Russian term meaning a group of three closely related things. Troika has three elements: a manned tank, an unmanned tank, and a low-cost, nonrecoverable rocket-propelled mini-air vehicle. The robot tank would be remote controlled with some semi-autonomous capability, and linked to the manned tank either through fiber optics or radio-frequency control. The unmanned tank stays in close proximity to the manned tank, providing escort and defensive-action missions. The RPV would be launched from the manned tank to furnish advance warning of enemy tanks and antitank weaponry. The advantage of Troika is that it allows a tank crew to command the firepower and mobility of two tanks and also have some aerial reconnaissance support. Troika would not be useful for massed tank maneuvers, but does seem to have merit for tank patrols in which the robot tank could take much of the risk away from the manned tanks' crews.

BOMBED-OUT ROBOTS

Teleoperators and hybrid versions with some preprogramming capability have already been in widespread use by the military and the police in dealing with terrorism and bombs. Britain led this development in building vehicles that could remotely deal with bombs set by the Irish Republican Army and Protestant extremist organizations. The Wheelbarrow is the most popular *explosive ordnance disposal* (EOD) robot in use throughout the world. It is outfitted with a T.V. monitor, a shotgun for explosive detonation, various booms and mechanisms, and tow ropes and cable. It was developed in a joint effort by the U.K. Military Vehicles Experimental Establishment (a Ministry of Defence research organization in Surrey) and the Royal Ordnance Corps; Morfax Ltd. received the license to make the machine and has sold some 500 Wheelbarrows to military and police organizations in some forty countries.

Hunter and Hadrian

In light of the Wheelbarrow's success, EOD robots are being manufactured by a number of other companies. Two other British-built robots, the Hunter, made by the SAS group of Beaconsfield and the Hadrian (Figure 3.5), built by Monitor Engineers of Wallsend, were present at the 1984

Figure 3.5. The Hadrian remote-controlled robot. (Courtesy of Monitor Engineering.)

Olympic Games held in Los Angeles, for stand-by use in coping with any terrorist bombs. The Hunter is a radio-controlled vehicle that has a unique TRAVAD auxiliary-drive system that quickly allows for the option of using tracks for surmounting obstacles, or wheels for greater speed on paved surfaces. Its actuators provide a maximum lift capacity of 220 lb (100 kg), and the telescopic arm can reach 13 ft (4 m) with an angle of elevation up to 87 degrees. It can mount either a single or a pair of robotic arms, with four movements in four places. The grip has a rotation of 360 degrees and its pincers can exert a grip pressure of up to 120 lb. The wrist has 160 degrees of action, and the elbow can flex 90 degrees. The arm includes a concealed five-shot semiautomatic 12-gauge shotgun, which is used in conjunction with a laser target designator that enables a shot to hit a 1-in. (2.54-cm) target at up to 50 yd (45 m). The Hunter can carry two synchronized pan-and-tilt cameras plus one that is independently controlled. Two of the cameras can also transmit sound that, when used in conjunction with other installed communications equipment, allows for conversations to be held in siege/hostage situations. The fully coded digital radio's pulse-control modulation is controlled by a microprocessor. If radio interference occurs, the Hunter will switch to a preprogrammed fail-safe hold position until the signal quality is regained.

Besides bomb disposal, the Hadrian robot is also used for mobile surveillance in prisons and other areas, in SWAT operations, and in hazardous waste areas. It has six drive wheels that allow it to climb and descend stairs, curbs, and steep embankments at speeds up to 3 mph (5 kph). Its gear-driven articulated arm is on a 360-degree rotating turret and can extend to a reach of 94 in. (2.4 m). At the end of the arm is a chain-driven lifting and manipulating claw with a wrist tilt of 180 degrees that can grasp and retrieve objects as well as turn door knobs and dials. A heavier claw capable of picking up 165 lb (75 g) can be quickly fitted in place of the manipulating claw. The Hadrian can place and detonate explosives via two separate, secure firing circuits, which are controlled from the command console. It also can carry a five-shot semiautomatic shotgun to disable the bomb's firing mechanism or clockwork.

U.S. Development of Ordnance-Disposal Robots

The U.S. Navy's Explosive Ordnance Disposal Technical Center (NAVEOTECHCEN) at Indian Head, Maryland, has an ongoing program to develop remote-control and teleoperated systems for neutralizing explosive devices. They have taken the lead for triservice research in this area. The navy-funded program under their direction is called ROVER (for Remotely Operated Vehicle for Emplacement and Reconnaissance). This land-and-ship-based system can be controlled either by radio or by cable, and has an on-board video camera that "sees" for the operator. The

ROVER's control functions are microprocessor-based. It has a three-degree-of-freedom manipulator for grabbing objects. SAMSON (for Semi-Autonomous Mobile System for Ordnance Neutralization) is an enhanced version of ROVER with a six-degree-of-freedom manipulator plus tool-changing capability and greater on-board sensory capability. The center at Indian Head has been funded by the U.S. Army for the development of two explosive ordnance-disposal robots. The first is named RON-D, which is a similar remote-controlled mobile platform to that of the ROVER, except that it has enhanced manipulator capabilities not found on the ROVER. The second funded robot is the ReCORM, which is a very small, one-man portable vehicle that is remote controlled and has a T.V. camera system.

A joint Israeli–U.S. corporate effort called 21st Century Sivan Ltd. has produced a series of different explosive-ordnance robots. The Israeli Army and security forces have ordered a number of the small TSR-50 remote-controlled tracked vehicles. The TSR 700 Wasp is a larger wheeled version, and the TSR 150 is a medium-sized robot that can be outfitted with wheels or tracks.

Remote-controlled robots are also being developed for use in other ordnance-removal tasks besides the disposal of terrorist bombs. The USAF 3207th Munitions Maintenance Explosives Ordnance Disposal Branch at Eglin Air Force Base, Florida, has built its own remote-controlled robot, which is designed to recover and diffuse armed, live warheads. The robot, nicknamed Robbie, picks up the live fuse and warhead and loads them into a sawing mechanism that cuts off the section of the warhead containing the fuse. While the saw is doing this, the robot moves away to a safe distance in case the bomb is accidentally detonated. Use of Robbie permits the air force to recover live munitions in order to diagnose the cause of failure.

The U.S. Navy's Heavy Equipment Remote-Control (HERC) system, built by Battelle Memorial Institute, Columbus, Ohio, was designed to clear explosive ordnance at firing and testing ranges. It consists of a control kit that can be mounted on a variety of heavy equipment. The control console can be hand-carried. To date, HERC has been tested on eight vehicle classes, ranging from a backhoe to a tank.

A major U.S. Air Force concern is the repair of airfield runways that have been cratered and then permeated with chemical munitions. People functioning in this environment, even if wearing NBC-protective clothing, would be operating at a much slower rate than usual, perhaps not quickly enough to get the airfield back into service when needed. The filling and repairing of the runway craters and the removal of bomblets on a chemically polluted airfield may only be possible with the employment of a vehicle such as the Rapid Runway-Repair Excavator being built by John Deere and Company (Figure 3.6). This vehicle, still in development, can be operated either from the cab or through remote control by the addition of a "black box." Under hazardous NBC conditions, the operators can be protected in a citadel

Figure 3.6. The Rapid Runway-Repair Excavator, which can either be manned or run by remote control through use of a "black box." (Courtesy of John Deere and Company.)

control station or within an armored vehicle at a safe distance while the robotized excavator is on the field. The John Deere Rapid Runway-Repair Excavator is being tested at the U.S. Air Force Research and Development Division of the Engineering and Services Center at Tyndall Air Force Base.

The U.S. Navy's Explosive Ordnance Disposal Technology Center is directing a similar program called the Remotely Operated Mobile Excavator (ROME), which is being built by the Foster Miller Company. The ROME vehicle is also designed primarily to provide a capability for quick access to unexploded ordnance embedded in airport runways.

Robots in Minefields

During World War II, the German army was the first to use robotic vehicles to sweep mine fields. These vehicles were inexpensive systems designed to sacrifice themselves in order to clear the mines. In the 1960s, Bernie Lumbert, an enterprising visionary, was unsuccessful in an attempt to sell a small robot that would detonate mines to the Pentagon. Lumbert maintains that the Thumper, which at the time would have cost only $500 a piece, could have prevented thousands of foot and leg injuries in Vietnam from booby traps, which cost the government about $50,000 per injury in order to evacuate the wounded soldier and to pay him benefits for the rest of his life. The U.S. Army has perfected this idea with the introduction of a vehicle that does not have to be sacrificed in order to clear minefields. The U.S. Army Tank–Automotive Command in Warren, Michigan, has completed development of their prototype of the remote-controlled mine-clearing vehicle called the ROBAT (which stands for Robotic Obstacle Breaching and Assault Tank). The ROBAT consists of two armored pods that are mounted on top of the base plate of a modified M60A3 tank. It is designed to propel explosives that detonate sequentially into enemy minefields, causing the mines to explode and thus clearing a path for the other vehicles. The ROBAT is guided through remote control across the path to cause any remaining mines to explode. While doing so, the ROBAT also marks the cleared lane with chemiluminescent light sticks. Robot Defense Systems designed a remote-control system for the ROBAT that consists of a T.V. camera and a fiber optic cable to carry the signal to a remote-control monitoring station. The army hopes to eventually have 142 ROBAT vehicles.

INDEPENDENT ROBOTS

Michael Brady and Patrick Winston, two robotics researchers at the Massachusetts Institute of Technology, have described intelligent robots as having "connected perception to action." The *autonomous* robot is considered "smart" because its computer or "brain," like a human's, acquires knowledge by perceiving its environment, which is done by collecting information from its "eyes and ears," the sensors. This connection to action is how decisions are derived, and the robot's reasoning ability comes from synthesizing new concepts out of the recently acquired information and out of its memory of stored knowledge.

The use of robotics in conjunction with artificial intelligence to create autonomous vehicles has captured the imagination of the public, as well as having caused controversy among military and civilian defense analysts. Autonomous robots have certain advantages over remote-controlled and teleoperated vehicles, which justifies the high expenditures and intensive research efforts in artificial intelligence and sensors. A teleoperated system replaces a relatively cheap infantry man with a more expensive robot technician. As long as there is a one-to-one correspondence between unmanned systems and their operators, overall manpower costs will not be reduced. Only when robots assume their own direction, without human involvement, can we save on manpower costs. According to the U.S. Census Bureau, by the year 2000, the segment of the population capable of serving the military (ages 18–23) will sharply decrease by 20%. The U.S. Labor Department has also projected an additional 2 million entry-level jobs by 1990. These two factors indicate that the military may be faced with increasing difficulties at recruiting good people from the more lucrative civilian job market. Autonomous robots can fill the gaps left by the shortage of skilled servicemen by performing certain specialized functions, including replacing the operation of teleoperated robots.

A major advantage of autonomous robots over teleoperated vehicles is in the command and control area. Remote-controlled and teleoperated vehicles require that the human operator receives the input from the robot's sensors, and transmits signals to guide its actions. The communications between the operator and robot are susceptible to jamming. Messages can also be intercepted and reveal the robot's and operator's location to the enemy. The exchange of information may also use up a lot of valuable time and slow down the robot's ability to perform its mission. An autonomous vehicle has the ability of making more rapid decisions, and is not dependent on communications with the operator. The crew manning the teleoperated system is likely to get fatigued over time. Its performance will vary with the quality of the human operator. The autonomous robot will only tire after its power source runs down. It can remain quiet and passively detect enemy movements while recharging itself. Another advantage for autonomous vehicles is that they may be able to coordinate more effectively and efficiently among each other, as to their separate courses of action, than could a manned operator performing control of several teleoperated systems, or as compared to how a number of human operators would coordinate among themselves.

There are several missions that are well-suited for autonomous robots. The deep-attack role behind the enemy's first echelon, to interrupt the flow of fuel and supplies to the front, is of primary importance to the U.S. Army's Air–Land 2000 doctrine, which emphasizes maneuverability over attrition warfare. Autonomous robots could effectively accomplish this by wreaking havoc on the enemy's supporting infrastructure without fear for their own safety. Nor would they be dependent on communications (which would have to be transmitted through the enemy's air space) that are likely to be jammed. The autonomous robot could help stymie the enemy's attack and buy the needed time to help mobilize forces to prepare for a counterattack.

Reconnaissance missions in areas affected by electromagnetic pulses (EMPs) resulting from detonated nuclear warheads can also be performed with greater certainty by autonomous robotic systems. The data links required by teleoperated systems are likely to be severely disrupted under these conditions.

The operators of teleoperated systems are vulnerable to nuclear, biological, and chemical weapons; their deaths would also knock the teleoperators out of commission. Autonomous vehicles are able to perform their mission regardless of the condition of the rest of the force structure.

A number of military leaders are quite fearful of losing direct control over their weapon systems, and therefore are negatively disposed towards autonomous vehicles. This is based on two beliefs: First, such systems may lessen the overall power and influence that manned operators would maintain in the military bureaucracies; second, and more realistic, is that by removing the man "out of the loop," it would make weapons tactically deficient by being unable to adapt to new circumstances unforeseen by the artificial intelligence (AI) programmers. The only way to respond to these fears is to point out that in the near future, and certainly in our lifetimes, autonomous vehicles will represent only a small minority of the weapon systems in the inventory. If autonomous vehicles become the prevalent weapons system in the long term, it is only because they would have by that time mastered Darwin's "survival of the fittest" theory by being able to rapidly adapt to changing situations.

ALV Program

The "granddaddy" of all the U.S. autonomous robot programs is the Autonomous Land Vehicle (ALV) program being jointly developed for the Defense Advanced Research Projects Agency (DARPA) and the U.S. Army Engineer Topographic Laboratories (Figure 3.7). Although it is designed as a vehicle that can perform reconnaissance and other missions on the battlefield, its primary purpose is not to furnish a near-term "smart" robot, but rather to serve as a demonstrator test bed in order to expedite our understanding of vision and image technology. Such information will be

Figure 3.7. Autonomous Land Vehicle traversing rugged terrain. (Courtesy of Martin-Marietta.)

necessary for successful development of the longer-termed autonomous vehicles that will emerge at the turn of the century. The ALV is one important component of DARPA's $600 million Strategic Computing Program, which is designed to give the United States a broad base of machine-intelligence technology, and to demonstrate its applicability to the development of modern weapons systems.

Under a $17 million, five-year contract, Martin-Marietta Denver Aerospace Inc. was given the primary responsibility for developing and integrating robotic technologies with advanced artificial intelligence architecture within the ALV's existing eight-wheeled all-terrain vehicle. This vehicle is totally self-contained by means of AI and advanced sensory-perception technologies. Lasers, video cameras, and radars feed data to a computer system, enabling the vehicle to react and determine its own movements. The contract calls for a demonstration of the developments achieved each year. In May 1985, the ALV was able to travel over a 0.6-mile (1-km) preset course on a paved road at a speed averaging 3 mph (4.8 kph). A vision-processing system pioneered by the University of Maryland analyzed a frame of imagery every 2.4 seconds. This enabled the ALV's on-board computer system, Digital Equipment Corporation's VAX 11/785, to interpret the data, and then to direct the robotic vehicle to the center of the road. In November of 1985, the ALV performed the same tasks at twice the speed.

By 1992, the ALV program is projected to demonstrate a vehicle capable of an on-and-off-road traveling speed of up to 60 mph (96.5 kph) and possessing the ability to avoid obstacles and to perform the remote-sensing capabilities required for battlefield reconnaissance. An emerging technology that will be researched in conjunction with the ALV is a five-color laser-scanner radar that can measure absorption and reflection at the five different wavelengths, in order to determine the nature of an obstacle.

Other Army Autonomous Robot Projects

Some of the remote-controlled and teleoperated robotic vehicles described earlier have the ability to be enhanced, with the proper software and sensor integration, in order to demonstrate autonomous mission capability. In another DARPA-funded program, which took place at Fort Lewis, Washington, in August 1985, Robot Defense Systems's PROWLER successfully demonstrated robotic artificial intelligence by negotiating around a nonlinear, contoured 500-ft (143-m) fence, making a 180-degree turnaround upon reaching the end, and retracing its original route. While performing this maneuver the PROWLER also practiced an NBC reconnaissance mission for detecting a simulated chemical agent with the use of the M841 government-furnished chemical detection device and alarm, which had been installed on the PROWLER.

The U.S. Army Tank–Automotive Command's (TACOM) Advanced Ground Vehicle Technology Program has demonstrated autonomous robotic vehicle technology capable of a route reconnaissance mission. Both FMC Corporation and General Dynamics Land Systems Division have been given funding to develop their own version of the Advanced Ground Vehicle. The FMC test vehicle is based on a tracked M113 personnel carrier outfitted with various sensors mounted on its front slope. The General Dynamics version (Figure 3.8) is a wheeled armor car with sensors mounted on mid and aft locations. Both vehicles utilize autonomous navigation breakthroughs pioneered by DARPA's ALV program. These prototype vehicles will be used to explore the robot's mobility and navigation capability, communications between the vehicle and a command center, the soldier–machine interface, and specific mission applications. In this last area, the Ft. Knox Armor Center is working with DARPA to conduct robotic combat vehicle mission analysis. This is aimed at determining what the user community wants to do with the robotic vehicles, what degree of autonomy is needed, and the cost and scheduling estimates for fielding such vehicles. The armor center is prioritizing missions according to which are deemed most essential for such a robot. These are then broken down into subtasks, enabling their researchers to address specific problem areas.

The U.S. Army has also developed conceptual designs for an autonomous

Figure 3.8. Artist's concept of General Dynamic's Advanced Ground Vehicle. (Courtesy of General Dynamics.)

medical evacuation robot named the Artificial Intelligence/Robotic Supplement to Medical Support (AI/RSMS). This concept envisions a mobile platform that would detect and recover up to six wounded soldiers on a battlefield, and then bring them to field hospitals or medical stations for treatment. This would lessen the risks to humans by reducing the need for medics and Medivac helicopters to enter fire-fight zones, and it would also recover the injured soldiers at a much faster rate, thus saving lives.

Robart I

The U.S. Navy's support of autonomous robotics research efforts, spearheaded by Lieutenant Commander Hobart Everett, serves as a fine example of how a military bureaucracy can develop an individual's interests and channel it, with the proper funding, into an extremely worthwhile program. Everett was an accomplished robotics hobbyist prior to enlisting as an officer in the navy. While at the Naval Postgraduate School from 1980 until 1982, Everett was engaged in developing an autonomous robot using current technology that would demonstrate the feasibility of such systems and point the way towards future developments. This 5-ft (1.5-m) prototype robot, named the Robart I, was designed to randomly patrol a building sensing for fire, smoke, flooding, toxic gases, and intrusions. It would follow up detection with a warning and an appropriate response.

Robart I was guided by a forward-looking ultrasonic ranging unit, a long-range near-infrared proximity detector that was positioned in a rotating head, ten short-range near-infrared proximity detectors, and tactile feelers and bumper switches for collision avoidance. In addition to this, Robart I had a unique homing device that would activate when the vehicle's battery needed recharging. When the voltage dropped below an acceptable level, the robot would signal via radio link to a beacon at its recharging station. A 75-W light bulb served as the beacon, which was tracked by an optical photocell array located on the robot's head. Only in 1 out of 200 dockings did the Robart I fail to reach its recharging station. The robot was powered by a 12-V 20-A·h battery that furnished it with ten hours of service. The recharge period required fourteen hours.

The Robart I was equipped with an infrared-body-heat sensor that could detect a person out to a distance of 50 ft (15.2 m). A near-infrared long-range proximity sensor with a parabolic reflecting collector allowed the system to detect the edges of an open doorway, which enabled it to enter and exit without any difficulty. A voice synthesizer gave the Robart I the ability to verbally warn of alarm conditions, and to report the status of its own internal mechanism and system configuration, as well as operational information. In the guard mode, the Robart would announce "Intruder, Intruder" when detecting a person breaking in, and would then ask the individual to leave the room. Under normal circumstances, it would greet individuals with a friendly "Hi" or "Hello."

The Robart I was able to randomly choose, from a preprogrammed set of sixteen routines, what to perform during any given time. This, however, was overrided when any of the sensors obtained some abnormal input. At this point, the standard software was interrupted, and the software designed for responding to the alarm would take over to meet a specific goal. This AI programming made the robot's behavior appear arbitrary. It moved directly ahead, unless confronted by an obstacle, at which point it would swerve to one side and continue on.

The Robart I's sensors interfaced with a 6502-based SYM-1 computer. The body moved on a triangular wheel base with the front wheel providing power and steering. It had a rotating head and an analog/digital (A/D) converter that furnished four bits of information to steer the wheel angle.

Robart II

In 1982, Everett and Anita Flynn, an M.I.T. master's degree program co-op student who was working at the Naval Surface Weapons Center, began development of an enhanced model, the Robart II (Figure 3.9). This version is 4 ft (1.2 m) high and 17 in. (43 cm) across at the base. The number of on-board computers has been increased from one to eight, which allows for some parallel processing and increased sensor usage for path planning, collision avoidance, and environmental awareness. It has six ultrasonic rangefinders, fifty near-infrared proximity detectors, and a long-range infrared rangefinder. Four infrared motion detectors are used to detect intruders up to 75 ft (22.9 m) away. Other sensors have improved abilities to detect smoke, fire, flooding, vibration, and toxic gas. An added self-diagnostic feature allows Robart II to request operator assistance when needed. It has two linear charge-coupled device (CCD) cameras, and a speech-recognition board that can recognize up to 256 speaker-trained words. Everett's success with the Robart series of robots resulted in his appointment as the director of the Office of Robotics and Autonomous Systems at the Naval Sea Systems Command.

Navy, Marine, and Air Force Projects

The Naval Surface Weapons Center is also developing a three-dimensional vision system with an extremely narrow depth of field that can isolate objects of interest from other clutter. A motorized lens system focuses back and forth at a frame rate of 30 Hz. This allows a three-dimensional model of the scene to be constructed from successive two-dimensional pictures. Glass optics and analog electronics are used for image processing, including feature extraction and clutter rejection. Unwanted details are filtered out. This vision system will be demonstrated on a low-silhouette battlefield-robot prototype under development at NSWC.

The Naval Ocean Systems Center (NOSC) at San Diego is developing the Ground Surveillance Robot (GSR) (Figure 3.10) with funding from the U.S. Marine Corps. This effort, headed by Scott Harmon, is geared toward demonstrating the ability of autonomous systems to perform certain Marine Corps missions. The GSR is built upon an M113 armored personnel carrier that houses fifteen on-board microprocessors that have 8.5 Mbytes of RAM. It also is equipped with roll and pitch angle sensors that monitor altitude, and a continuous-wave (CW) radar doppler that furnishes ground speed. Other subsystems include a satellite navigation receiver, an acoustic proximity sensor, a laser rangefinder, and a magnetic compass. Its vision system consists of a high-resolution camera that is mounted on a compass-controlled platform having pan, tilt, and vertical extension motions. The

Figure 3.9. Robart II displaying internal circuitry. (Courtesy of U.S. Navy.)

Figure 3.10. The Ground Surveillance Robot. (Courtesy of U.S. Navy.)

combination of camera and laser rangefinder enables the GSR to accurately model surrounding terrain. The GSR can, without human direction, track and follow other vehicles across obstacles. Near-term plans are to demonstrate its ability to traverse across simple obstacles using its vision, rangefinder, and proximity sensors. The GSR program's long-term goals are to examine its ability to circumnavigate through unknown terrain. The Marine Corps has not yet specified any of the actual missions that would be performed by an operational version of the GSR.

The U.S. Air Force has embarked on a program aimed at developing a robot with artificial intelligence capability by the year 2000 that could be used to refuel, rearm, and perform minor repairs to aircraft. A 4-ft (1.2-m), 150-lb (68-kg) prototype vehicle nicknamed Marvin is being developed at the Harry G. Armstrong Aerospace Medical Research Laboratory at Wright-Patterson Air Force Base, Ohio. Marvin is powered by two 24-V direct-drive motors and moves on small concealed wheels. It can turn its head, move its arms, and grasp objects. The follow-up advanced version of the Marvin is viewed by the Aeronautical Systems Division at Wright-Patterson as being especially useful in performing repair and maintenance work at airfields under hazardous conditions, such as those existing during chemical and biological weapons attacks.

WALKING TO WAR

The term "all-terrain vehicle" is a misnomer. There is no land vehicle in any army's inventory that can climb mountains, jump across ravines, travel through swampy marshes, and perform in all of nature's diverse settings. Military wheeled vehicles can journey on only 30% of the Earth's land surface, and a little under 50% cannot be traveled on by any tracked vehicles. Due to these limitations, many military forces throughout the world still rely on "beasts of burden," including horses, mules, and camels, to transport supplies over rugged terrain. As recently as the early 1980s, U.S. Army leadership tried to convince certain state National Guard organizations to reintroduce mules into special logistics units. Needless to say, few soldiers would be tempted to be in "Today's Army" because they were motivated to tend mules. The sociological obstacles combined with the lack of any political lobbying effort supporting the use of mules prevented the notion from being forced onto the guardsmen. There are other serious drawbacks in using animals. They are limited in the loads they can carry and they require a great deal of food, water, and sometimes tender loving care. They may spook in a fire fight, and are, of course, also vulnerable to physical injuries.

Chapter 2 described how developing a machine that can accomplish animal-like locomotion has intrigued mankind for many years. The development of a practical legged machine would start a new age in off-road vehicles. Two major types of terrain remain, for the most part, impassable to wheeled and tracked machines. These are the very soft, swampy-type areas and the mountainous regions with large, irregular depressions and projections. Tracked vehicles may traverse some swampy areas, but not without a real loss in efficiency, because a lot of energy is used to push the mud out of the way, in order to leave a path. In contrast, a legged vehicle would step through the mud, moving very little on the surface. It would be like riding a bicycle through soft sand instead of walking over it. The effort required for cycling as opposed to walking is much greater. A legged machine is also more energy efficient because of its ability to step across or climb over obstacles, even those of greater size than itself, rather than having to circumnavigate around, as required by wheeled vehicles. It can also furnish a relatively smooth ride over rough ground by varying the length of its legs to match the undulation of the ground.

One major design problem facing developers of walking machines is how to incorporate energy efficiency into the leg design. Energy efficiency in a walking vehicle is hard to achieve because the vehicle must support its weight as well as propel itself. The energy exerted by each leg is classified into three work groups: (a) the positive energy mode, used for movement;

(b) the isometric energy mode, used for maintaining balance and holding the machine in a certain position; and (c) the negative energy mode, used for the stopping movement. Building a legged vehicle that uses energy in the positive mode, but that also receives energy in the negative work mode, is the goal of several research organizations, and would be considered a significant milestone.

Most animals, including humans, use what is called "dynamically stable" walking. This requires moving forward to an off-balance stance and then catching yourself with the forward placement of the other leg. In a fast walk or run, the dynamically stable walking animal is constantly off balance. Although this is the most efficient form of walking, it also requires complex control systems that are difficult to develop, even with today's computer technology. Methods are being researched that may some day allow dynamic walking by machines to be realized, but for the meantime, legged machines will primarily rely on a "statically stable" walk. The crawl used by most insects and spiders is a form of statically stable walking. This type of locomotion requires at least four legs, with three being stationary at all times to achieve balance. Many of the walking machines currently being developed have six legs. This allows for a tripod gait where three legs move at one time, leaving three in a stationary position. The legs alternate with each step. A tripod gait is extremely stable because it is very adaptable to the roughness of the terrain. Six legs, as opposed to four, reduces the weight on each leg. Eight legs or more may needlessly complicate the drive mechanism and control equipment.

The vehicle's ability to interact with its environment is of great importance to maintaining balance. Because most of this interaction is through its "feet," a sensing mechanism located there is a necessity. Being able to determine obstacles and ground proximity greatly reduces the need for any operator input, and therefore advances the technology closer to reaching autonomous operation. Many different designs are being tested, with varied success. Most use a foot with a combination of sensitive detection wires called "whiskers," and sole switches to obtain a rudimentary ability to sense varied terrain.

Ohio State University's Hexapod and ASV

A group at Ohio State University has been performing walking-machine research since the late 1960s. The OSU scientists have concentrated on the fundamentals of static and dynamic stability, gait planning, foothold coordination, and force coordination associated with both four- and six-legged vehicles. A six-legged, electrically tethered vehicle called the OSU Hexapod has served as the primary testbed for much of their work. Under a $4 million grant from the Defense Advanced Research Projects Agency

(DARPA), OSU built a much larger machine named the Adaptive Suspension Vehicle (ASV). This vehicle, which was described in Chapter 2, can climb high obstacles while carrying one passenger. The ASV has a scanning mechanism that senses obstacles and then sends instructions to a series of computers that control the six legs and ensure a coordinated movement in the right direction. The driver controls the general operation of the vehicle, such as speed and direction, much in the same manner as does the rider of a horse. In neither case does the driver or rider concern himself with the individual leg movements.

In 1985, DARPA announced a program to merge the Adaptive Suspension Vehicle with the autonomous vehicle research being conducted by Martin-Marietta Corporation. This project will cost $50 million and will run through 1990. Surprisingly, a legged vehicle may need less artificial intelligence than a wheeled vehicle because a wheeled vehicle must examine alternative routes to circumnavigate around an obstacle, whereas a legged vehicle can simply climb over it.

The Carnegie-Mellon Project

Besides OSU's Adaptive Suspension Vehicle, DARPA has also funded legged-machine research at Carnegie-Mellon University (CMU). Their vehicle, designed by I. E. Sutherland, has an unusual hip actuator arrangement. Horizontal movement is obtained from two cylinders that are mounted in a V-shape above the leg. When one valve shortens, the other lengthens. A third actuator serves as the knee to produce sideways movement. The CMU hexapod is 8.2 ft (2.5 m) long and weighs 1,764 lb (800 kg). It is powered by a 13-kW gasoline engine driving four variable-displacement pumps. This vehicle actually beat out the ASV as the first man-carrying computer-controlled walking machine. The original prototype has been scrapped, although CMU researchers continue to conduct theoretical studies from its test results.

Odetics's Functionoid

Odetics Inc. of Anaheim, California, has produced a legged-robot concept, termed a 'functionoid." The first functionoid prototype has been dubbed the Odex 1 (Figure 3.11) and has been designed to perform a wide range of military tasks, which are made possible through its attributes of mobility, profile changeability, self-contained power, strength, and stability. In appearance, the Odex 1 resembles an octopus, except having six instead of eight legs. These legs, or "articulators," are evenly spaced around its cylindrical body. A video camera is mounted on top and is protected by a "fish bowl" clear dome. The camera can rotate 360 degrees,

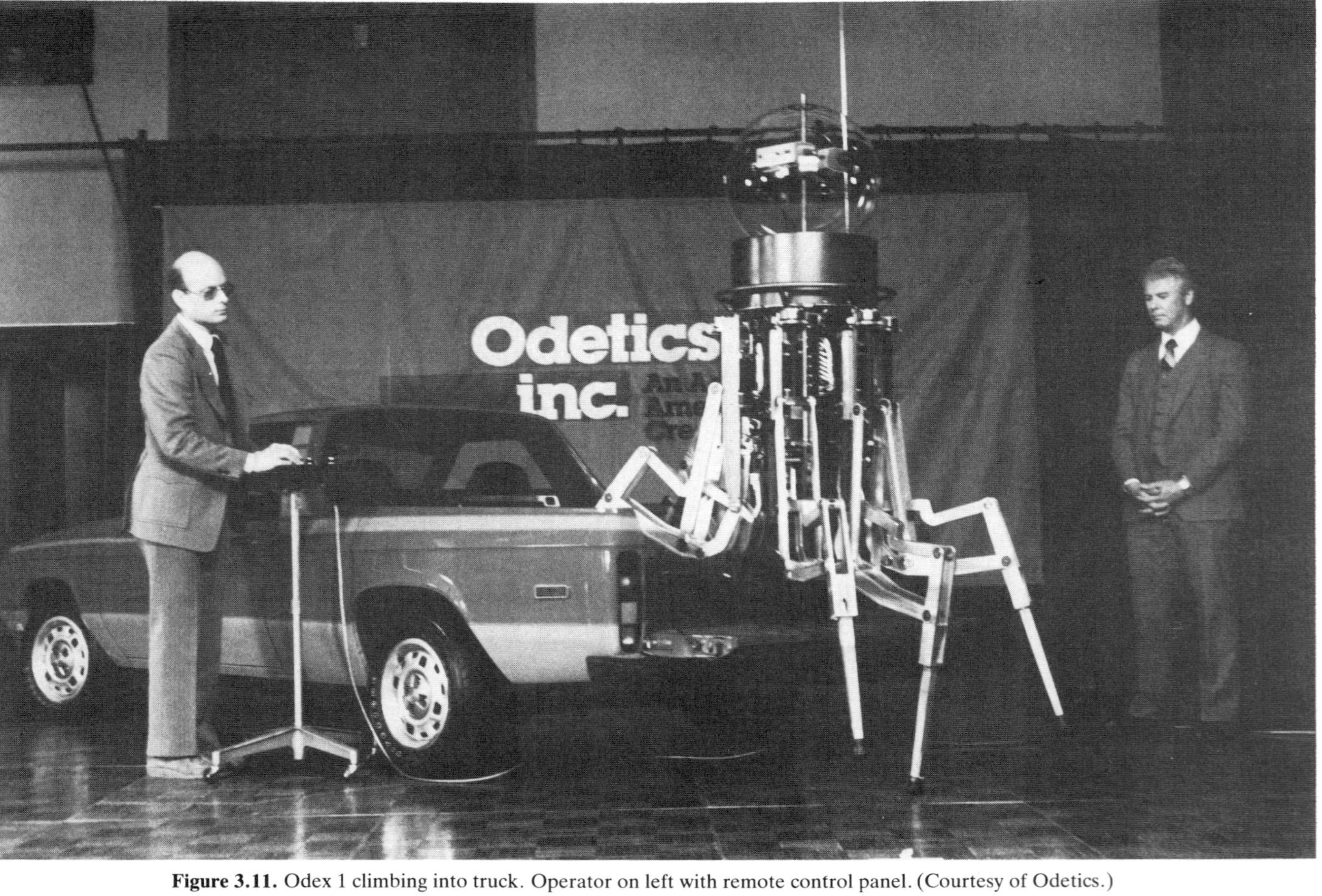

Figure 3.11. Odex 1 climbing into truck. Operator on left with remote control panel. (Courtesy of Odetics.)

allowing the Odex 1 to be totally omnidirectional. Motive force is provided by three electric motors for each articulator. One moves the leg vertically, one extends the leg, and the third swings it. Power is supplied by a self-contained 24-V 25-A·h aircraft battery that has a one-hour operating time without charging. The Odex 1 walks by using a tripod gait in which three legs move at a time and then alternate with the other three, which had been stationary. In this manner three legs are always on the ground, providing maximum stability and better speed. The Odex 1 can travel 8 mph (12.9 kph), which is a very fast walk by human standards. It can change direction in midstride and rotate on the move. Seven microprocessors (one per articulator and one as a central unit) control the Odex 1. It is currently operated through remote control, by use of a joystick. Furnishing it with artificial intelligence so that it may operate autonomously on very global orders, is under research by an in-house project group at Odetics. The Odex 1 has the added ability to change its profile. Height can be varied from a low of 36 in. (9.4 cm) to a high of 72 in. (182.9 cm). This is done by having the robot either stand or squat. The width can be altered by positioning its legs 105 in. (266.7 cm) wide for maximum stability or a narrow 21 in. (53.3 cm) for moving through a tight doorway. The Odex 1 can operate in "human areas" such as buildings, stairways, and elevators, and can step over objects as high as 33 in. (83.8 cm).

Odetics considers the functionoids' unique strength-to-weight ratio as a crucial asset; therefore much of their research has been devoted toward perfecting this capability. The design of the Odex 1's articulators allows it, while in a stationary position, to lift loads of 2,100 lb (1,053.4 kg), which is over five times greater than its own weight of 370 lb (168 kg). It can carry 900 lb (408.6 kg) while operating at a normal walking speed.

The creation of the Odex 1 has led to numerous research contracts for Odetics from both the government and private sector. The U.S. Army's Human Engineering Laboratory (HEL), Aberdeen Proving Ground, Maryland, has contracted with Odetics to develop a design for a "high payload-to-weight manipulator structure." When completed, a manipulator arm weighing less than 500 lb (186.5 kg) will be able to lift a 500-lb (186.5-kg) payload while extended 15 ft (4.6 m). A larger version will be capable of lifting as much as 4,000 lb (1,492 kg) at a 25-ft (7.6-m) reach. Eventually this manipulator will be used in forward-area material-handling tasks and be mounted on top of a variety of vehicles.

Other Odetics Projects

The U.S. Navy is interested in a functionoid-type vehicle for ship-board fire fighting. The Naval Surface Weapons Center at White Oak, Maryland, has provided funding to Odetics to examine the use of a legged machine as a teleoperated fire-fighting hose-delivery system. Fighting fires on board aircraft carriers is an extremely hazardous duty, regardless of whether during peace or wartime. It is a complicated task involving fully fueled aircraft parked closely together, which can result in a chain-reaction fire. Detonations may scatter burning debris in all directions, which may contain live ordnance, creating further problems and explosions. The Odetics study defined the basic fire-fighting requirements for quickly extinguishing or containing fires on a flight deck without risking human life.

How robotics might be incorporated into physical security operations is being examined by the U.S. Defense Nuclear Agency. As part of this research, Odetics was awarded a contract worth over $450,000 to examine applicable legged robotic systems and to define their technical specifications.

Box 3.2 gives an example of the possible uses of a "walking machine" in the twenty-first century.

EXAMPLE OF ASV OR FUNCTIONOID USE IN THE TWENTY-FIRST CENTURY

Box 3.2 The future land-battle robot, resembling a large bug, will be fully autonomous during the conduct of its mission. Its six legs will duplicate and even improve upon the obstacle-traversing capability of man. In addition to its all-terrain maneuverability, it will have many features designed to allow it to defend itself and to carry out offensive actions or intelligence-gathering duties. A typical reconnaissance mission might consist of locating any enemy defenses in a particular geographic area (including mine fields, booby traps, tank ditches, and the enemy forces). Moving only at night, the hexapod would not be easily detectable with its low profile and small heat signature. If the enemy is discovered, the robot could remain motionless for days, gathering intelligence with video cameras, microphones, radar, and infrared sensors. It could also have chemical and radiation sensors on board for use in NBC environments. All of the gathered data could either be stored or transmitted through encrypted signals. Weaponry on an autonomous robot could include nonlethal arms such as strobe lights powerful enough to blind a human, and sonic disrupters that would leave a man paralyzed and in great pain for hours. Lethal weaponry could include any type of small arms, as well as machine guns, grenade launchers, antitank rockets, or a combination of same, depending on the mission. The robot would be easily transported by truck or armored vehicle. It could be airdropped behind enemy lines to carry out reconnaissance and sabotage missions.

Japanese, French, and Russian Research

Researchers at the Tokyo Institute of Technology, under the direction of Dr. Shigeo Hirose, have also closely examined the insect world to aid in their development of legged vehicles. Although using a four-legged prototype vehicle, unlike an arachnid's complement of eight legs, these Japanese researchers have developed an innovative design that duplicates the body positioning of a spider (Figure 3.12). The center of gravity is lower than that of other legged vehicles, which increases stability, allows the device to move faster, and incurs less energy loss for the support of body weight during walking. These electrically powered prototypes use a wire-and-pulley mechanism. They have eight tactile sensors and are tethered to a microcomputer.

The Japanese corporation, Komatsu Ltd. has developed a legged vehicle for underwater construction. The ReCUS (Remotely Controled Underwater Surveyor) is a large 29-ton vehicle that is 26.2 ft (8 m) long, 16.4 ft (5 m) wide, and 19.8 ft (6 m) high. The ReCUS consists of an inner frame that has four vertically telescoping legs. The inner frame slides longitudinally on the bed of the outer frame. Its walk is achieved by simply standing alternately on the outer and inner leg sets, thus accomplishing an alternating tetrapod gait. A PDP-11/03 computer on board the mothership controls the ReCUS movements. The ReCUS is designed for performing construction tasks down to depths of 230 ft (70 m). Although primarily conceived of as a tool for civilian purposes, a vehicle such as this would also be extremely useful in military underwater construction, including the setting up of special acoustic sensors, the recovery of objects, and the building of underwater submarine bases.

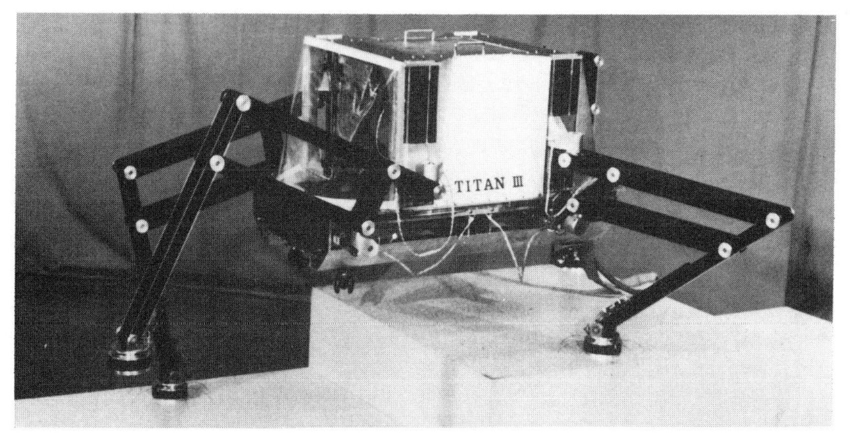

Figure 3.12. Titan III robot resembling arachnid body positioning. (Courtesy of Tokyo Institute of Technology.)

Researchers at the University of Paris have built an electric hexapod. Instead of using a third actuator to serve as a knee in each leg, they have made the legs laterally flexible, which enables them to bend during a turn.

The USSR's Moscow State University Institute of Mechanics has developed a six-legged hexapod similar to the ASV. This walking machine is equipped with a vision system for obstacle avoidance, as well as a sophisticated mechanical interaction with central circuits enabling the vehicle to maintain equilibrium. A Leningrad researcher named Ignatiev has worked on a hexapod whose legs are radially situated about a central vertical axis, in a manner that resembles the Odex 1. It is quite likely that the Soviet prototype vehicles' offspring, as with U.S. robots, are intended for military applications.

Wheeled and tracked vehicles are here to stay for the foreseeable future, but there is a warfare niche that only legged vehicles can truly fill. Combat in the twenty-first century will most likely include the use of walking machines resembling giant insects. The future battlefield may furnish a second connotation for the term "army ants."

Bipedal Robots

Since the early 1970s, researchers at several Japanese universities have been experimenting with bipedal robots that mimic human walking. The Wabot, built by Waseda University, has features that correspond to the human anatomy, including legs, arms, and head-mounted sensors. This hydraulically powered robot walks with a gait in which the statically stable states alternate with unstable ones. The University of Tokyo has developed several prototypes of dynamically stable bipedal robots. Each of the BIPER 4 robot's legs has a hip roll, ankle pitch, and knee joint, and is connected to the other leg by a hip joint. Its actuators are ac motors, and it has contact sensors located in each plastic foot.

Resembling a human does not necessarily make a better robot. Anthropomorphic robots are only beneficial in military environments that require a humanoid-type body to operate weapons or machinery. For example, in order to fly a jet fighter, a robot would need to have manipulators that could duplicate what the pilot accomplishes with his extremities within the same operating space. An easier approach to accomplishing a specific task is to "robotize" the vehicle. The human body is not the optimum configuration for performing most specialized tasks, whereas the robot can be designed for the specific environment it will operate in. As we require robots to perform a greater variety of generalized tasks that are normally performed by humans, there will be a need to develop robots that resemble the human anatomy.

HOPPING AND CRAWLING

If rolling on wheels or walking on legs cannot get the robot to a particular objective, then perhaps it can get there by hopping or crawling. Scientists at Carnegie-Mellon University's Robotics Institute have been conducting experimental work on a one-legged hopper that bounces like a pogo stick. An umbilical cable supplies compressed air, hydraulic oil, and control signals to the hopper. Their test vehicle has successfully hopped around a square path that included some sharp turns. There are two possible applications for this mechanism. The first is that this type of locomotion may prove useful in a low-gravity environment such as that on the moon. There have been conceptual studies that envision giant pogo-stick-like vehicles propelling astronauts across the moon. The other application for this research is developing a four-legged running machine. A trot-like gait has been obtained in computer simulations in which two of the pogo-stick-type legs thrust in phase, with the differential thrust being used to regulate the vehicle's body pitch. This research is aimed at enabling legged vehicles to achieve speeds comparable to those of a race horse, or to what wheeled vehicles could reach on level surfaces.

Professor Shigeo Hirose of the Tokyo Institute of Technology has been experimenting with flexible structures whose movements resemble those of a snake or a centipede. Robotic systems of this type are called active-cord mechanisms (ACM), and they can gently conform to objects of any shape (Figure 3.13). One of Hirose's ACMs consisted of a series of short segments connected by hinge or ball joints, with a single azimuth actuator at each joint. By thrusting backwards against walls with undulating waves of the body, the ACM was able to travel between obstacles and through a labyrinth, much like a snake would do. Such mechanisms may furnish the military with some unique applications. Crawling vehicles would be difficult to detect and could be useful in performing special operations, such as the sabotage of oil pipelines by crawling through the pipes. They may be used as expendable weapons by sneaking their way into sensitive installations and vehicles, and then exploding themselves upon reaching a vulnerable location.

SUPPORTIVE ROBOTS

Besides being weapons platforms and ordnance-removal vehicles, robotic vehicles can also serve as important mechanisms for support missions and other noncombat functions under hazardous conditions. These include replenishment and refurbishment, munitions handling, and transportation.

Standard Manufacturing Corp. has developed the Remote Transport Platform (RTP) to provide for the remote-controlled, high-speed transport

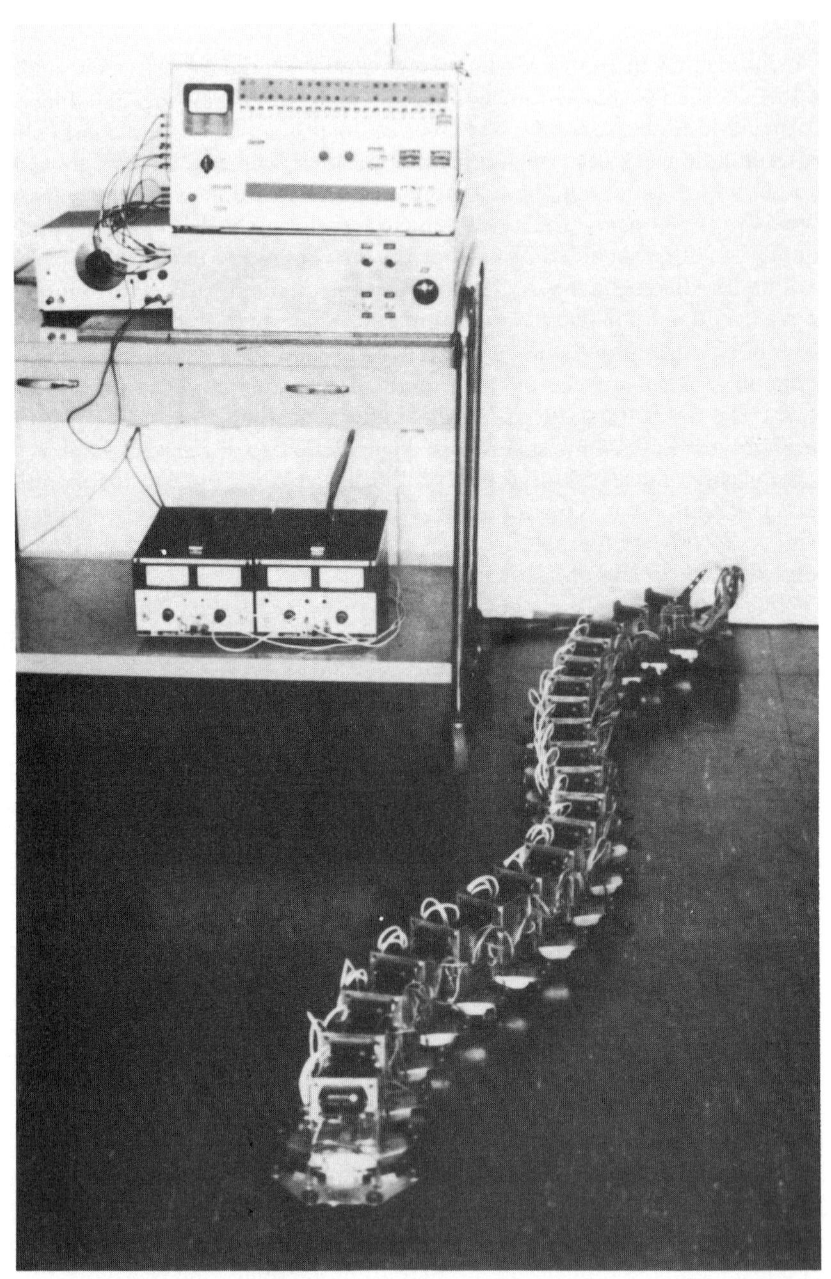

Figure 3.13. Active cord mechanism with numerous feet, providing a likeness to a centipede. (Courtesy of Tokyo Institute of Technology.)

of needed supplies over rough terrain. This vehicle has slightly different dimensions than their Fire Ant assault vehicle, although it can reach a comparable speed of up to 85 mph (138 kph), and also has a very low profile, making it difficult for the enemy to intercept.

Battelle Corporation laboratories have produced what they call the Radio or Computer Operated Mobile Platform (ROCOMP) (Figure 3.14). This mobile platform, which could be used to carry military supplies, resulted from an in-house program to establish a technological base for research in robotics hardware. The ROCOMP is designed to perform useful work in a variety of military and civil environments. It has demonstrated its capability

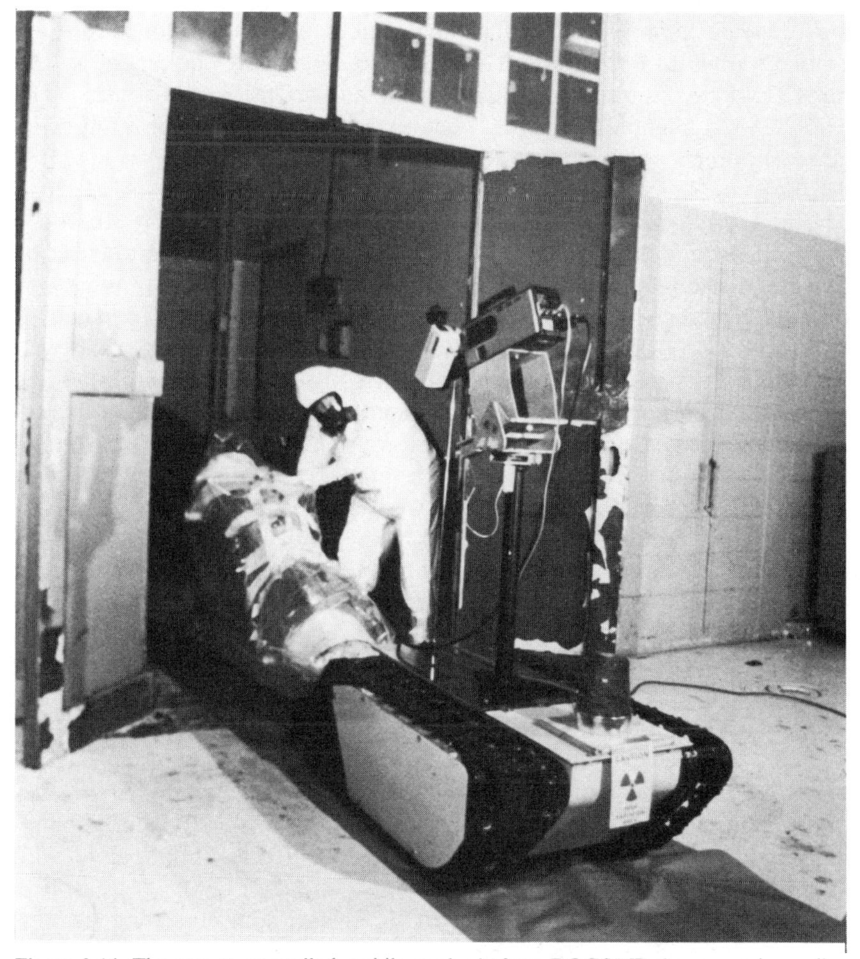

Figure 3.14. The remote-controlied mobile work platform ROCOMP shown moving radioactive material. (Courtesy of Battelle.)

for jobs in nuclear power facilities, chemical plants, and in security roles and fire-fighting duties. The ROCOMP's statistics:

- 18 in. (46 cm) high;
- 28 in. (71 cm) wide;
- 54 in. (136 cm) long; and
- weight is 250 lb (113.5 kg).

It is driven by tracks, which allow for operating up and down ramps as well as stairs. Its narrow width permits usage in narrow halls and doorways. The ROCOMP can be controlled through an umbilical cord, by radio instructions, or be maneuvered via a programmed itinerary. Battelle is working on dead-reckoning or obstacle-avoidance navigation for projected semiautonomous operations. They are also developing equipment for the ROCOMP to monitor the environment and to recognize objects. The ROCOMP's manipulating arm, when extended, is capable of lifting 50-lb (23-kg) objects and when retracted can carry items weighing up to 200 lb (91 kg).

The U.S. Army Human Engineering Laboratory (HEL) at Aberdeen Proving Ground has been examining the use of robotic systems to facilitate the rapid handling of ammunition. They have built a 25-ft (7.6-m) articulated arm. The huge appendage has six axes, flexible joints, and can lift and carry 3,800 lb (1,725 kg). HEL has also been experimenting with attaching Unimation 400 and Unimation PUMA 760 robots onto a flat-bed trailer to ascertain whether they could handle 8-in. (20-cm) artillery projectiles, each weighing 200 lb (91 kg). This program, called RALS (for Robotic Ammunition Loading System), was developed at Tooele Army Depot in Utah. RALS proved successful both without sensors (by using fixed positions for the robots), and then by applying sensors and having the robots search for the projectiles. An extension of this testing, called the Robotics Test-Bed Program (RTB), used robots with sensors to locate and position a 3,800-lb (1,725-kg) pallet of ammunition. The National Bureau of Standards supported the testing by developing a special robotic control system that permits high-speed, instant real-time sensor-data processing on a new Unimation 9000 computer-controlled robot. In June 1985, the U.S. Army released a request for proposal to design, fabricate, and demonstrate within 20 months a prototype heavy-lift pallet-handling robot system that involves transferring industrial robotics to field handling. This field-material-handling robot (FMHR) would operate in a work-cell configuration along with other automated equipment.

In order to quickly douse aircraft carrier fires without subjecting the crew to fire-fighting injuries, the U.S. Naval Surface Weapons Center has been examining how robotic systems can be used for on-board fire-fighting applications. Under the direction of project manager Mary Lacey, remote-controlled vehicles have been designed, tested, and critiqued as to how reliably and safely they perform operations in the hazardous environment associated with shipboard fires. The remote-controlled vehicle also served as a platform to test advanced sensors used for locating the fire and aiding the operator in maneuvering the vehicle. The platform used for these vehicles is a hydrostatic-drive front-end loader powered by an internal-combustion engine and equipped with a trainable nozzle. The operator can control the nozzle and other functions using a hand-held console up to distances of 500 ft (152 m) away from the vehicle. The navy hopes to eventually procure 100 fire-fighting robots to be located on aircraft carriers. These vehicles will probably encompass an all-tracked system capable of operating at speeds of 10 mph (16 kph) and controlled through a fiber-optic link.

TARGETS ON THE MOVE

The use of remotely controlled targets allows tank gunners, missile operators, and attack aircraft and helicopter pilots to practice firing on mobile systems. In the late 1970s, the defense division of Brunswick Corporation had built land-target systems that resembled the Soviet T-64 tank. These vehicles were developed on behalf of the U.S. Air Force Advanced Systems Directorate's A-10 Program Office at Wright-Patterson. The models cost $8,000 a piece and consisted of a U.S Army Mule all-terrain vehicle, which was covered with styrofoam and canvas. The land-target systems were used by A-10 attack aircraft at Nellis AFB to practice strafing runs. They were much less expensive as compared to the cost of retrofitting outmoded armored vehicles into targets.

The Spyder, built by the British Company Flight Refuelling Ltd., is a remotely controlled vehicle that resembles an unmanned jeep. This 6-ft, 6-in. (2-m) long vehicle with a height of 3 ft (1 m) is primarily a target-training aid for tank, antitank, ground-attack aircraft, and guided-weapon crews and operators. It can be operated at out-of-sight distances of up to 3.1 mi (5 km). The Spyder can be employed in other roles, such as tank decoys, for surveillance, mine detection, and sweeping, and for operations in hazardous areas, through use of its remote-control radio/video system and carrying capacity. The U.S. Army has recently robotized M113 armored personnel carriers for use as noncooperative targets.

In the next chapter as we focus on remotely piloted vehicles, we will see that the U.S. Army has not been the only military service that has experienced tension between proponents of manned systems and the advocates of robotics. At the conclusion of World War II, the leaders of what was soon to be known formally as the U.S. Air Force recognized that robots would become part of their future. These visionaries' successors, however, have not always been as open to robotics, and the battle lines were drawn.

NOTES

1. *R & D Plan For Army Applications of AI (Artificial Intelligence)/Robotics*. SRI International, Menlo Park, CA, 1983. This study listed the numerous combat and support applications for robotics to perform army missions.

4 Current Operational Use of and Developments in RPVs

The morning after the Japanese surrendered to U.S. forces in 1945, U.S. Army Air Forces General Hap Arnold assembled his staff and gave them a talk about the future of air warfare. What he had to say made his audience sit up and take notice, although many had not yet recovered from the all-night celebration following V-J Day. Arnold stated,

> We have just won a war with a lot of heroes flying around in planes. The next war may be fought by airplanes with no men in them at all. It certainly will be fought with planes so far superior to those we have now that there will be no basis for comparison. Take everything you've learned about aviation in war, throw it out of the window, and let's go to work on tomorrow's aviation. It will be different from anything the world has ever seen.[1]

The wars fought since Arnold's prophesy have been fought primarily with manned aircraft, although unmanned aerial vehicles have been playing an increasing role in performing various essential missions. The need to furnish ground-force commanders with timely intelligence coupled with the high cost of manned aircraft and their vulnerability to modern air defenses has led to a great deal of enthusiasm in certain defense quarters for the widespread use of RPVs to conduct reconnaissance and other missions. Meanwhile, advocates of manned aircraft have viewed RPVs as a drain on the dollars that would be devoted to manned reconnaissance aircraft.[2] A bias against unmanned systems by officers whose careers and outlook have been shaped on the pilot's ability to fly also exists. Nevertheless the technological, economic, and political factors involved would indicate an increasingly greater reliance on unmanned vehicles. Current and projected applications are being considered for RPVs, not because they are being shunned by the flyers, but because they can in fact perform certain missions more effectively and economically than can manned vehicles. The developments as revealed in this chapter may turn Arnold's vision into a belated reality.

RIGHT ON TARGET

A natural niche for remotely piloted vehicles is that of aerial targets. Few operators of manned vehicles have been willing to serve in this capacity, at least not in noncombat missions. The growth of airborne weaponry's range, speed, altitude, and destructive power has increased exponentially over the past several decades. Aerial targets designed to simulate aircraft and missile performance have also had to become more complex and sophisticated. Because targets are of lesser priority when it comes to the budgetary process, such systems also must remain economical. The challenge therefore is to enhance target capability without dramatically increasing cost.

There are several reasons for using aerial targets. They aid in the development of a new weapon system by validating its ability to perform the intended mission, such as shooting down the simulated aircraft. Whether or not a current system can meet a new threat can also be tested. Aerial targets provide a mechanism with which to check maintainability, reliability, and the quality assurance of a given weapons system. They are also used for air crew and surface-to-air missile crew training.

Aircraft and missile performance and tactics are mimicked with both subscale and full-scale targets. Subscale targets are usually cheaper to develop, maintain, and operate. They are used to simulate airspeed, altitude characteristics, simple maneuvers, and infrared (IR) and radar signatures. Full-sized targets are often missiles or manned aircraft that have been converted into drones. Not all aerial target testing involves the actual impact of a missile warhead or shell. Aerial target testing and training can involve just tracking and missile launch alone, or missile flight through warhead fusing. In this way, targets can be flown a number of times.

Chukar 11 and the NV-144

Northrop Corporation, which had acquired Radioplane, is the manufacturer of the Chukar 11 MQM-74C turbojet-powered subscale target, the most widely used aerial target in the world. Since 1968, when the U.S. Navy began its use, it has been added to the inventories of 12 countries. This high-subsonic-speed target is used for gunnery and missile crew training, as well as weapons-system evaluation. The enhanced Chukar 11 version is launched from tactical aircraft; it has programmed flight profiles, an expanded payload, and an automatic low-altitude control down to 30 ft (9.2 m).

Northrop's latest aerial target is the NV-144, which is designed to carry large, heavy payloads through the high-subsonic-flight environment. The NV-144 incorporates digital avionics and advanced plastic technology for its airframe materials.

Firebee 232 and Firebolt

The Firebee series of aerial targets were introduced in 1951 by Teledyne Ryan Aeronautical. They have been continuously upgraded; the newest version is quite different from the original prototype. The latest Firebee, model 232, incorporates a new avionics suite that features a three-axis flight controller. The vehicle's computer allows it to perform complex maneuver sequences patterned after the capabilities of manned aircraft. This 2,500-lb (1,125-kg) vehicle can travel over 600 nmi (1,062 km), is capable of maneuvering at 6 Gs, and can fly up to mach 0.97 from altitudes of 600,000 ft (18,280 m) down to 10 ft (3.18 m). The Firebee can be equipped with a towed radar transponder and infrared simulators so that live warheads can be used without destroying the vehicle. It is being used to simulate sea-skimming missiles for the U.S. Navy.

Teledyne Ryan has developed their first digital-computer-controlled target, the AQM-81A Firebolt. Funding for this rocket-powered target is provided by U.S. Air Force Armaments Test Laboratories at Eglin Air Force Base. It can fly at speeds from mach 1.2 to 4.0 at altitudes up to 100,000 ft (30,460 m). No other aerial vehicle in existence can fly as fast and high at level flight. The Firebolt is designed to simulate both air-to-air and surface-to-air missiles. The U.S. Navy is also using the system to simulate cruise missile attacks on their Aegis fleet defense system. The Firebolt is launched from an F-4, at which point it ignites and flies a preprogrammed mission. At the end of the cruise, it dives to a lower altitude and at a certain point a recovery chute is deployed. Recoveries can be made in midair by helicopter, or it can be picked up from the sea. This vehicle is 17 ft long (5.29 m), and has a diameter of 13 in. (33 cm) and a gross weight of approximately 12,000 lb (5,400 kg). The target has a 16-bit microprocessor that controls the continuously variable rocket engine, manages flight controls, stores and implements the mission programs, and controls the payload, navigation, and communications. An active radar-augmentation system and an autonomous scoring system make up the payload. The radar-augmentation system emits a radar cross section that can vary, depending on the desired simulated threat. It uses a doppler radar scoring system that measures miss distance. Each Firebolt is built to last twenty flights, but is considered cost effective if it performs three flights.

Beech Aircraft's RPVs

Beech Aircraft Corporation is another large supplier of high-performance aerial targets for the U.S. military. Their AQM-37A target missile has been in use since 1963 by the U.S. Navy. It can operate at supersonic and subsonic flight. Recent versions incorporate modified wings and a solid-state autopilot. Beech Aircraft's MQM-107 uses a JATO booster rocket. It has radio-command guidance for maneuvering and recovery, but can be preprogrammed during flight. Beech's latest target missile is the BQM-126A, which is primarily designed to test fleet air-defense units. It represents the threat of aircraft or cruise missile attack and can fly from sea level to an altitude of over 40,000 ft (12,186 m) at speeds of mach 0.9, with a 7-G turn capability. It can be launched from land, the deck of a ship, or the wing of an aircraft.

SLAT

The U.S. Navy has awarded Martin-Marietta Orlando a contract to develop the Supersonic Low Altitude Target (SLAT), which has the requirement of carrying a complex threat-replication payload at mach 2.5 from 50 mi (80.4 km) out at an altitude of 30 ft (9.2 m). This performance is required for simulating the Soviet SS-N-22 cruise missiles. The SLAT engine is to be a ramjet with an integrally mounted engine. Because SLAT is being fired at our own ships, and not the enemy's, it must navigate more reliably and accurately during its midcourse flight. Unlike missiles, it has no end-course requirement.

Full-scale Targets

An economic way of producing high-performance targets is by modifying retired squadron aircraft into full-scale targets. Those used by the U.S. Air Force include the QF-106 (see Figure 4.1), PQM F-102, and QF-100 aircraft, which are drone conversions of surplus F-106, F-102, and F-100 fighter jets. These vehicles serve as a realistic and predictive representation of enemy aircraft for advanced weapon testing and combat crew training. They can be launched in the air by a towing aircraft. The newer models, however, are also capable of automatic takeoffs and landings. Sperry Corporation, one of the "founding fathers" of remotely piloted vehicle development, received a $16.7 million contract in June 1986 to furnish the U.S. Air Force with six QF-106 full-scale aerial target vehicles. They also have options for converting 184 more F-106 manned interceptors into the drone versions.

Figure 4.1. QF-106 drone aircraft. Van contains Vega Corporation remote-control equipment. (Courtesy of Vega.)

Russian and European Target RPVs

The Soviets have had several aerial target programs, but very little is known about them. The UR-1 is a short-range subscale target drone designed to operate from 65,400 ft (10,000 m) to 130,800 ft (20,000 m). It is launched from an aircraft and is remotely controlled. The full-scale Mandrate target is a modified version of the high-altitude reconnaissance Yak-25 fighter. The existing fuselage was married to a new straight wing. The vehicle is remotely controlled and used for surface-to-air missile testing.

A number of other aerial targets are under development or in full-scale production throughout the world. The French firm Aerospatiale has been an innovator in target development. They have introduced their second-generation C22 target drone, which is a successor to the CT20. It can fly at mach 0.95, is capable of maneuver-induced loads in excess of 6 Gs, and simulates hostile aircraft and missiles operating between 50 and 45,930 ft (15 and 14,000 m). It has a flight endurance of two hours and can be recovered over land and water by parachute. A naval version of the C22 was first exhibited in October 1984. It is being marketed to countries that do not have firing ranges on land.

The Italian firm Meteor has developed their Mirach series of target drones. The Mirach-70 is a conventional monoplane with a 70-hp engine. The Mirach-100 is a small turbojet-powered target, with a one-hour endurance. Both the 70 and 100 models are built under license in Argentina. The Mirach-300 is similar to the 100, except that it has larger engines and a two-hour endurance. The British Flight Refueling ASAT/Falconet target drone was built for the British Army. It has a small turbojet, with an endurance of up to ninety minutes.

The Chinese-made Changcheng B-2 is a 16-hp, piston-engined weapons-training target. India's Aeronautical Development Establishment first flew its ship/ground-launched variable-speed subsonic target, the PTA, in 1984. It is powered by a single turbojet and has an endurance of one hour.

RECON RPVs

Furnishing RPVs with interchangeable payloads allows these versatile vehicles to perform a multitude of missions, including surveillance, electronic countermeasures, target practice, and the delivery of ordnance. It is the reconnaissance mission, however, that most RPVs have been designed for and the bulk of their operational planning is directed towards that function. The growing need for supplying field commanders with timely intelligence, combined with the escalating cost and vulnerability of manned systems to surface-to-air missiles, has led to a worldwide willingness to invest in reconnaissance drones.

Short-range RPVs

The Aquila

The U.S. Department of Defense's Joint Resources Board developed a "road map" in 1985 with which to guide remotely piloted vehicle development. The U.S. Army was given responsibility for overseeing all new short-range RPV efforts. The Lockheed Aquila has been the army's primary reconnaissance RPV developmental program. Numerous technical and managerial setbacks have, however, delayed the program, causing a loss of some political support in Congress. As a temporary "Band-aid" solution, the army decided to purchase off-the-shelf general-purpose RPVs to provide some capability until most of the problems associated with the Aquila were "ironed out." Upon the Aquila's introduction into the force structure, the less-capable drones may be given to the reserves.

The developmental experiences of the Israeli Scout and the U.S. Army Aquila provide an interesting contrast in design and acquisition philosophies and practices (Figure 4.2). Both programs were initiated shortly after the 1973 Middle East War. The Israelis had an urgent need to field a mini-RPV as quickly as possible. Their near-term operational needs typically won out over long-range research benefits, and thus they were able to introduce the Scout and the Mastiff RPVs into their force structure by 1978. For several years the Israelis were able to train and operationally experiment with their RPVs, a fact that no doubt contributed to their effective use against the Syrians in 1982.

By contrast, it took over twelve years of development until the Aquila

RPV was ready to be fielded. The 1976 Aquila design was a relatively cheap, $100,000-a-piece RPV. The original delta-wing body was made from a difficult to see nylon derivative. It was powered by a modified go-cart engine. The current Aquila is a much more sophisticated RPV with such features as a jam-proof data link, a much smaller radar signature, and a laser-beam system to pinpoint targets at night. Some critics of the Aquila program have accused it of "gold plating." A strong case, however, has been made by U.S. Army officials that Aquila would be used in the European theater where electronic countermeasures and air defense systems employed by the Soviets would be much more threatening and pervasive than what existed in the Middle East. They maintain that in order to survive in this environment, the Aquila needed vastly superior technology than that incorporated in the Israeli RPVs. Its cost, however, has risen eight times higher than originally estimated, making it quite unpopular with Congress. The army might have been better off introducing a less-capable Aquila early on, and then incrementally adding technological enhancements, instead of trying to create the "perfect" vehicle prior to actual development.

After a concerted effort by Lockheed to correct deficiencies ranging from

Figure 4.2. The stealth-configured Aquila. (Courtesy of Lockheed.)

acquiring, tracking, and automatic recovery to reliability and durability, the army decided in February 1986 to go ahead with full-scale development. The airframe is fabricated from preimpregnated Kevlar epoxy material. The Aquila weighs 260 lb (97 kg), is less than 7 ft (2.1 m) long, and has a wingspan of less than 13 ft (3.9 m). With enough fuel for three-hour missions, the Aquila can carry a payload package of 60 lb (27.2 kg). The mission payload consists of a daylight black-and-white T.V. camera, laser rangefinder designator, stabilized optics, and a sophisticated electronic sensor chassis that is mechanically coupled via a rotating optical joint to a moving turret assembly. The ground controller can select from three different T.V. camera fields of view. The laser rangefinder designator is optically aligned with the T.V. camera and can be used to accurately track and measure distances to targets or to designate targets for laser-guided munitions. The optics and laser maintain their lock on stationary and moving targets regardless of the RPV's maneuvers. A spread-spectrum, antijam data link provides the command uplink to the air vehicle, and status and video downlink communication to the ground-control station.

The Aquila is launched from a hydraulic catapult on a truck. During launch, the RPV's propeller hub is engaged by the hydraulic starter motor. A piston, acting through a cable-sheave stroke multiplier, drives the assembly along the rail, accelerating the air vehicle to launch speed. The recovery subsystem involves a vertical Dacron net mounted on a truck. As the Aquila enters the net, it is decelerated by extending lines attached to kinetic-energy absorption devices.

Pioneer 1

The U.S. Navy and Marine Corps have begun to take delivery of their Pioneer 1 short-range RPV, which is manufactured by the Israeli Mazlat Corporation in conjunction with the Baltimore-based AAI Corporation. Mazlat is a joint venture between Israel Aircraft Industries (IAI) and Tadiran. As can be expected, the Pioneer 1 technology borrowed heavily from the operationally proven Scout and Mastiff programs. The 400-lb (181-kg) Pioneer 1 is a high-winged monoplane made from metal and fiberglass. It is propelled by the West German Sachs 26-hp rear-mounted engine. The RPV has a wingspan of 16 ft (4.8 m) and can carry its heaviest payload of 100 lb (45 kg) to a maximum altitude of 15,000 ft (4,570 m). It has a very long flight endurance time of seven hours. A ground-based observer and an aircraft operator control the RPV with equipment produced by an IAI subsidiary. The ground-control station is extremely compact, and the display system can be carried on a soldier's back pack, in a Jeep, or even in a tank. Takeoff can either be accomplished from a launcher or from a short runway under the Pioneer 1's own power. It can perform a runway landing on its own wheels or be retrieved by net.

The Skyeye

In 1984, the U.S. Army purchased a squadron of the Skyeye R4E-40 RPVs, built by Developmental Sciences Astronics Division of Lear Siegler, Inc. (Figure 4.3). These were operationally tested and used in Central America near the site of U.S. and Honduran military maneuvers. Although one Skyeye was lost during the Central American operations, the U.S. Army has been pleased with its overall performance and is considering additional purchases to complement the Aquila RPV force structure. The Royal Thai Air Force has been using a squadron of the lower-powered Skyeye R4E-30s since 1982. Egypt is strongly considering the purchase of up to thirty Skyeyes.

The Skyeye is at the large end of the mini-RPV category, with a maximum launch weight of 530 lb (241 kg). Because of its large wingspan of 17.6 ft (5.36 m) and length of 12.2 ft (3.75 m), it is marketed as a versatile aerial truck that does not need miniaturized custom payloads and so can save on expense. It can carry 250 lb (114 kg) of fuel and payload for missions of up to eight hours in duration.

The Skyeye can be flown manually with a simple autopilot or in a programmed mode. When programmed, this RPV factors information such as heading, altitude, airspeed, and sensor commands that allows it to fly out of radio contact for limited times. It then records imagery and flies back into radio contact for further instructions. The airframe is constructed of Kevlar,

Figure 4.3. The Skyeye, used by the U.S. Army in Central America. (Courtesy of Lear Siegler.)

glass fibers, and graphite composites. Approximately 8 lb (3.6 kg) of radar-absorbing material is added to reduce radar reflections. Its aspect-radar cross section is not much greater than that of the Aquila. The Skyeye requires a six-person ground crew. The command control center and catapult launchers are mounted on trucks. The Skyeye puts down on its shock-absorbing skid for landings, which require a 400-ft (122-m) field length. In emergencies a parachute can be deployed.

Other Mini-RPVs

In 1985, the British Ministry of Defence selected GEC Avionics Ltd. and Flight Refueling Ltd. to produce the British army's new Phoenix remotely piloted surveillance system. This is a small fixed-wing, low-cost RPV that can furnish reliable, real-time pictures and data. It will be able to carry approximately a 100-lb (45.3-kg) payload with a flight endurance of four hours. Control command signals and data between the RPV and the ground station will be sent through a secure data link. The British army's battlefield artillery target engagement system will incorporate the Phoenix as an integral component to the system's overall functioning. The airframe is made from parts of sandwich composite construction that is designed to have low radar, infrared, and acoustic signatures to make it hard to detect. It utilizes modular construction, including plug-in wires, which allows the Phoenix to be easily assembled by soldiers with little training. The Phoenix is pneumatically launched, and recovered by parachute in an inverted position. The landing impact is absorbed by a crushable dorsal fuel tank.

Pacific Aerosystems, Inc. of San Diego, which was founded by officials of the Italian firm Meteor, has developed the Heron 26 mini-RPV. The Heron 26 is an all-composite (carbon–graphite) derivative of Meteor's Mirach-20 Pelican RPV, which has been purchased by the Italian army and navy. The Heron 26 consists of a high, fuselage-mounted wing that supports twin tailbooms. It houses a pusher-configured engine. The RPV is 12.9 ft (3.9 m) long and has a height of 3.8 ft (1.2 m). Standard 13.1-ft (4.4-m) wings or special high-altitude 20.7-ft (6.3-m) wings can be fitted to the Heron's fuselage. Sensor packages are either carried in the RPV's nose, or suspended from a turret platform below the fuselage. The Heron 26 can be remotely controlled or can be flown through preprogrammed instructions. This RPV uses an expendable takeoff booster rocket for launch and is recovered by a parachute that deploys from the fuselage. It has a normal mission endurance of five hours with a 75-lb (28-kg) payload and the high-altitude version can reach up to 20,000 ft (6,093 m).

The U.S. firm E-Systems (Melpar Division) has developed a reconnaissance mini named the E-90. It was developed from the E-75 harassment drone, and has a 13-hp two-stroke engine that enables the RPV to carry a 20-lb (9-kg) payload for a flight of three hours in duration.

In 1985, the Israelis shot down a small Syrian reconnaissance drone built by the Soviets. This mini-RPV, called the DR-3, is similar in appearance to the Skyeye and had twin booms and swept wings. This vehicle had a fixed (nongimballed) television camera. A newer Soviet model is expected to have a stabilized, steerable camera with zoom lens.

A number of other mini-RPVs are being developed around the world in countries as diverse as Japan, Brazil, China, and South Africa. The previously mentioned mini-RPVs, however, are representative of the designs and capabilities offered by these small reconnaissance unmanned aerial vehicles.

Mid-range RPVs

U.S. Research

In examining proposals for mid-range RPVs, the U.S. Navy has been looking for a vehicle that can conduct reconnaissance at high subsonic speeds while acquiring high-resolution imagery. They also want the vehicle to have the ability to detect and identify targets, and transmit data in real time and in a jamming environment. The U.S. Air Force signed a memorandum of understanding with the navy to jointly develop and procure such a system. The navy has the lead for building the vehicle itself, whereas the air force is tasked with developing an electro-optical sensor package for the RPV. One likely candidate for the mid-range RPV is Northrop's NV-144. This vehicle can carry out either preprogrammed missions or remote-controlled ones. It is 19.5 ft (5.9 m) long and 20 in. (50 cm) in diameter. The NV-144 has a maximum altitude of 52,000 ft (15,842 m), and a two-hour endurance flight time.

The Beech Aircraft Corporation Raider is another possible mid-range RPV. Derived from Beech's 1,090-lb (407-kg) MQM-107B missile target, the Raider is designed to carry electronic radar jammers, flares, and chaff, and reconnaissance, radar enhancement, and enemy weapon-systems evaluation equipment. It can fly for up to 2.6 hours.

European and Soviet Research

In 1986, West Germany purchased eleven complete Canadair CL-289 launch and recovery reconnaissance systems, which consists of a total of 200 RPVs. Dornier will participate with Canadair in the manufacture of these vehicles. The French have also purchased two CL-289 systems, comprising forty flight vehicles. The CL-289 is launched from a mobile system by a rocket booster and then continues flying on a preprogrammed mission. A turbojet propels the drone at log level in a quasi-terrain-following mode that is controlled by the RPV's doppler radar. The RPV transmits photographic and infrared line-scan imagery via its data-transmission link. After reaching its target, the CL-289 returns to its home base. The flight is terminated when the vehicle intersects the guidance beams of the homing beacon, at which time the turboengine is switched off and parachutes are deployed. Landing bags are also inflated to protect the vehicle and its sensors. West German officials estimate that the eleven CL-289 systems they purchased will provide a reconnaissance capability equal to sixty McDonnell Douglas RF-4E manned reconnaissance aircraft.

Up until the mid-1970s, the Yastreb was the primary Soviet reconnaissance drone. This was a supersonic, turbojet-powered RPV that was adapted from the T-4A strategic cruise missile. Relying on data obtained from the remains of the U.S.-made Ryan Firebee RPVs that crashed in Vietnam, the Soviets moved towards the development of miniaturized electronics, which allowed them to build smaller RPVs. The RPV versions of the smaller Shaddock/Sepal (SS-N-3/SS-C-1) vehicles replaced the Yastreb, which were in turn surpassed with the entry of the RPV version of the Sandbox (SS-N-12), which was twice as fast as the Shaddock. The Soviet BL-10, which is still in operational evaluation as an air-launched cruise missile, is a probable candidate for a future reconnaissance RPV.

Long-endurance RPVs

There has been a great deal of conceptual research and prototyping in the area of high-altitude, remotely piloted surveillance craft. Teledyne Ryan has proposed a high-altitude, long-endurance (HALE) aircraft named the Spirit, which is designed for electronic intelligence collection, communication relay, targeting applications, sonobuoy monitoring for fleet antisubmarine warfare, and long-range weather monitoring. This RPV has a wingspan of 85 ft (26 m) and a fuselage length of 40 ft (12 m). It is designed to cruise at 50,000 ft (15,232 m) with a 300-lb (112-kg) payload and remain on station for eighty hours.

The U.S. Navy has expressed interest in high-altitude surveillance drones that can furnish ships at sea with long-endurance, extended-area coverage and early warning. Boeing Electronics Company has built a very large experimental unmanned aircraft at its Moses Lake, Washington, facility. Although the program is very sensitive, Boeing officials acknowledged that this vehicle had the potential for a variety of missions, including reconnaissance, radio relay, and border patrol. Boeing has experience in high-altitude RPVs, as they previously developed in the late 1970s the prototype Strategic YQM-94A Compass Cope high-altitude RPV, which was later cancelled. The U.S. Defense Advanced Research Projects Agency is conducting research on a long-term program named Amber that is aimed at developing autonomous RPVs with flight endurance times a magnitude longer than existing vehicles (Figure 4.4).

Figure 4.4. Artist's concept of a solar-powered autonomous RPV. (Courtesy of AUVS archives.)

DECOYS

Many reconnaissance RPVs can be reconfigured with special electronic countermeasures equipment that can emit signals mimicking those produced by manned aircraft or even naval vessels. In this fashion, the enemy can be fooled into believing that you are in a different location from where you actually are. Radar systems can be tricked into locking onto the decoy RPV, thereby giving away their own SAM site frequencies, which provides the necessary information for antiradiation missiles to home in on their target. RPV decoys can also be used to attract enemy missiles away from their intended target, thereby sacrificing the unmanned system for the good of the ship.

A few drones were primarily designed for a decoy mission. One such vehicle, introduced in 1975 by the Brunswick Corporation Defense Division, was an air-launched tactical decoy drone named Samson. Although the U.S. Air Force tested the decoy from a F-4 at Eglin Air Force Base, Florida, neither the air force nor the navy opted to buy any, due to a lack of a formalized requirement. This is a "Catch 22" argument designed to protect the services from being influenced by outside companies to purchase equipment the leadership really does not want.

The innovative Israelis, however, were quite willing to test the new technology and eventually purchased numerous Samson systems. In 1982, the Israelis used these 300-lb (130-kg) mach 0.9 RPVs during their conflict with Syria in the Bekaa Valley of Lebanon. The Samsons were used effectively to emit electronic signals, mimicking F-4 aircraft, which tricked Syrian SAM-6 air-defense missile sites into flipping on their active radars to engage the apparent intruders. The Israelis were then able to launch radar-seeking missiles from their real aircraft to destroy the SAM sites. The following year the U.S. Navy lost three aircraft attacking the same SAM sites, which quickly changed their attitude towards the Samson. In April 1985, Secretary of the Navy John Lehman and Israeli Defense Minister Yitzak Rabin signed a joint agreement to develop and produce a missile that some analysts believe to be a newer version of the Samson drone. Congress appropriated money for the navy to buy an advanced version of the Samson for the 1986 fiscal year. This RPV, known as the Tactical Air Launched Decoy (TALD), is also built by Brunswick. The TALD is designed to emulate a strike aircraft to the sophisticated antiaircraft systems that are expected to be encountered in future combat missions.

ASSAULT AND SUICIDE RPVs

Remotely piloted vehicles can go on the offensive and "sting" with the equivalent lethality accorded to manned systems. The Israelis proved with their American-made Chukar RPVs that they could be used as a platform from which to launch missiles in actual warfare conditions. Numerous conceptual and prototype RPV developments have been designed for defense suppression and penetration missions, including the delivery of ordnance such as fragbombs and incendiaries, use as a platform from which to launch guided and unguided missiles, and even equipping drones with gun pods from which to shoot bullets at a target. Reusable RPVs such as the Development Sciences, Inc. Skyeye R4E-40 can carry unguided rockets or guided missiles against a variety of targets. Rotec Engineering, Inc. has proposed outfitting the remotely piloted version of their Panther 2t ultralight aircraft with Stinger missiles and rockets tipped with laser-following sensors. This extremely cheap RPV costs only $15,000 for the airframe, engine, and actuator, and can stay in flight for up to two hours. West German firm Dornier has been working on a class of drones that would be carried near a target by a manned attack aircraft. The drone would be dropped, and then guide itself over the target, dispensing munitions.

South African firm National Dynamics has developed their Eyrie strike mini-RPV. This vehicle has a unique rhomboid wing configuration and low-observable features. It carries up to four 2.75-in. (6.9-cm) rockets and with external ordnance has an endurance of 7.2 hours and a range of 99 mi (160 km).

Expendable RPVs and Targets

The idea of delivering munitions to a target by means of an expendable RPV has recently become the most popular approach for unmanned attack drones. In April, 1986, Donald N. Fredericksen, director of tactical warfare programs for the U.S. undersecretary of defense for research and engineering, spoke of a number of U.S. RPV programs that act as lethal "kamikaze" drones. Such vehicles, armed with warheads, are programmed to locate and destroy enemy vehicles, radar, and other hardware that emit a radio frequency. Rather than be used as a platform, these harassment drones search for specific frequencies and then follow the beam to its source, exploding on impact.

General Dynamic's Pomona Division had a joint program in the 1970s with the West German firm VFW to produce a low-cost expendable minidrone for the U.S. Air Force and the West Germans. It was designed to destroy enemy air-defense radar. The Locust (for low cost) vehicle was a good idea that unfortunately was cancelled when the Germans pulled out of the program.

In 1979, the Boeing Military Airplane Company continued where the Locust left off, by developing their own expendable drone. The Boeing Robotic Air Vehicle (BRAVE-200) is a 265-lb (120-kg) drone that can be operated and maintained by a two-man crew. Designed primarily as a expendable minidrone whose missions include destroying an enemy's radars, jammers, and other emitters, it can also perform electronic countermeasures and surveillance. The basic vehicle can be outfitted with interchangeable seekers, payloads, and software, thus allowing changes in mission profiles and adjustment to enemy countermeasures. On a typical mission the vehicle would climb to an altitude of 8,200–11,500 ft (2,500–3,500 m) and proceed to the target area. There it would loiter until the emitter is activated, at which point it fixes in and dives at the target, destroying it with a small conventional warhead.

The West German military has reinstituted an attack minidrone program known as the KZO. There are several competitors, including Boeing's Brave-200 and Dornier's proposed Kleindrohne Anti-Radar (KDAR) vehicle. Each drone would cost between $40,000 and $100,000 (U.S.). Dornier is also examining equipping the KDAR with different sensors for attacking tanks and other targets.

The West German firm MBB has been developing an autonomous drone, PAD, designed for use by the West German army to destroy enemy tanks, armored artillery, and helicopters (see Figure 4.5). The PAD drones are housed in a container that serves as storage, transport, preparation, and launch unit. Each container, which holds twenty PADs and can be mounted even on nonmilitary trucks, serves as its own launch platform. The launch is facilitated with help from boosters that detach themselves from the drone when the rocket fuel is expended. Upon departure, the canister wings on the attack drone are unfolded. The PAD is preprogrammed for several hours of flight in the search mode. Upon detecting a target, the PAD dives in at a steep angle. Its integrated terminal-guidance sensor directs the vehicle with high accuracy; the warhead is designed to destroy hardened targets. MBB is

Figure 4.5. TUCAN experimental attack drone being launched from ground rail. Was predecessor of PAD. (Courtesy of MBB GmbH.)

developing a version of the PAD for the KZO antiradar mission utilizing a different sensor package.

William C. McCorkle, director of the U.S. Army Missile Laboratory in Huntsville, Alabama, has been trying to drum up interest in a remotely piloted attack missile named the FOG-M (Fiber-Optic Guided Missile) since 1974. McCorkle in part developed this rocket-powered attack RPV in his own basement using off-the-shelf parts. The FOG-M is launched from a boxlike case. Upon launch, stubby wings sprout from the vehicle's body, and optical fiber resembling monofilament fishing line is spooled from the back of the rocket. The FOG-M's nose houses a T.V. camera equipped with a zoom lens that transmits a T.V. image through the fiber optic to the operator, who sets the corresponding controls that steer the attack drone into the target. McCorkle had great difficulty at first in marketing his pet project, which was unable to compete against established programs. His luck changed, however, with the successful demonstration of the FOG-M, in which an army infantry corporal with a few hours of training successfully maneuvered the attack drone through the windshield of a target helicopter located 6 mi (9.7 km) away. This excellent performance, combined with McCorkle's claim that the weapons could be produced for as little as $20,000 a piece, resulted in a lot of new advocates for the FOG-M. Pentagon interest has propelled the FOG-M up into the ranks of the well-funded established programs.

RPVs Against Manned Aircraft

For a number of years researchers have "toyed" with the idea of developing unmanned fighter drones to take on manned aircraft. In 1972, a U.S. Navy fighter pilot, sitting on the ground, remotely controlled a modified Firebee drone engaged in a dogfight with a manned F-4 Phantom. The RPV was able to execute 6-G turns without losing altitude. It evaded the Phantom's Sparrow and Sidewinder missiles and scored several simulated "kills" against its opponent.

The success of this and other such demonstrations, however, was never translated into a program to develop unmanned fighters; that is, until now. The natural bias of the "manned" pilot community has been challenged by engineering and defense policy decision makers. Francis X. Hurley, chief of mobility technology for the Engineering Sciences Division of the U.S. Army's Research Office, told an April 1986 meeting of the Association for Unmanned Vehicle Systems Washington, DC, chapter that the research office was studying the use of pilotless aircraft squadrons to gain air superiority above the forward edge of the battlefield. Sophisticated computers were envisioned as controlling and navigating the "mini-dogfighters" to attack enemy aircraft in a predetermined zone. Hurley made the case that technological advances in aeronautical performance are outpacing the ability of pilots to cope with them. Even pilots wearing G-suits will black out at 4–5 Gs, whereas the RPVs can maneuver to their greater structural limits. The autonomous dogfighter RPV could be programmed to attack anything moving in a predetermined area, or could be equipped with an identification friend-or-foe sensor system. Air-to-air missiles such as the AIM-9 Sidewinder could be fired from the RPV.

AGARD and the SOARFLY

In a similar research effort, the North Atlantic Treaty Organization (NATO) AGARD group has been studying the idea of introducing an unmanned air-superiority aircraft sometime after the year 2000. The RPV would house on-board artificial intelligence systems, replacing the need for a human pilot; however, major decisions could still be made by a human ground controller. The craft could supplement or replace the use of manned aircraft in high-risk environments. This study is also examining how cost-effective such a system would be.

A possible candidate for AGARD might be the conceptual model of a robotic vehicle the U.K. computer company Scicon Ltd. introduced at the 1986 British Army Equipment Exhibition. The 25-lb (11-kg), 5-ft (1.5-m)-long conceptual model is of a futuristic unmanned observation and attack aircraft. The vehicle, named SOARFLY (Scicon Observation and Attack Robot), is launched from a tube and features an X-wing design that can be used for helicopter flight when rotating, or jet flight when the X-wing is locked into a fixed position. The SOARFLY is propelled by an electric engine powered by a lithium cell, and in the helicopter mode the X-wing rotates by gas jets located at the wing tips. It climbs to a predesignated operational altitude to fly reconnaissance missions, and operates independently of ground control while searching for enemy activity with its radar and infrared spectrum sensors. The vehicle analyzes sightings and identifies enemy equipment. Data are encrypted and transmitted back to intelligence headquarters when something is worthwhile to report. A small charge underneath its nose converts the SOARFLY into an attack drone when it spots an enemy target worth sacrificing itself for. The SOARFLY is designed with a range of 100 mi (161 km) and a top speed of 100 mph (161 kph).

Cruise Missiles

The cruise missile is the most politically sensitive of the robotic airborne vehicles, primarily because it is unmanned, nuclear, and expensive. Most U.S. research is directed toward enhancements of current systems. The U.S Navy has developed a new variant of their General Dynamic-built Tomahawk land-attack missile. Designated the TLAM-D, it carries up to 220

BLU-97B submunitions in packs that are ejected to each side of the missile's flight path. An on-board computer controls the number of submunitions dropped on each target. This weapon is effective against aircraft on the ground, including those in shelters or revetments. The U.S. Navy has also produced a vertical-launch system for the Tomahawk that saves a great deal of space, allowing for the placement of more missiles, or for use on smaller combatants.

The U.S. Air Force is considering avionics improvements to their Boeing-built AGM-86B cruise missile. This would include adding penetration aids such as chaff, decoys, and flares to get the vehicle through Soviet air defenses.

The General Dynamics Convair Division in San Diego, California, is developing an advanced cruise missile (ACM) that incorporates stealth technologies, including a special geometric shape, composites, and an engine producing a low-infrared signature. As many as 1,300 ACMs will be produced, beginning in late 1989. Other U.S. cruise missiles have been conceptualized as supersonic capable vehicles.

The Soviets have been developing sea- and ground-launched versions of their AS-15, which is a small, air-launched, subsonic, low-altitude cruise missile resembling the Tomahawk. The sea-launched version, the SS NX-21, is small enough to be launched from torpedo tubes. The SSC-X-4 land version is a mobile system designed to support Eurasian theater operations. A larger cruise missile, the SS-NX-24, has been flight tested from a specially converted Yankee-class nuclear-powered cruise missile attack submarine. Although probably designed for nuclear warheads, the Soviet cruise missiles may be accurate enough to use as conventional warheads, thus posing a significant nonnuclear threat to high-value targets such as airfields and radar sites.

Smart Bombs

The U.S. Air Force has been trying to elevate the "intelligence" of its "smart bombs" by pressing development of a new computerized "brilliant bomb" that can find its own target without pilot guidance. The 2,000-lb (906-kg) Autonomous Guided Bomb, which is under development at the Armament Division of Eglin Air Force Base, carries its own computerized images of targets and searches for corresponding readings of the terrain with its infrared sensors. It can then attack the target while the manned aircraft is away from any potential air-defense systems. Each bomb is projected to cost approximately $100,000. The expense, however, seems justified if it can alleviate the difficulty inherent in the delivery of smart bombs in which the pilot has to get virtually on top of the target in order to guide his bomb on a laser beam of light to reach its objective. One U.S. Air Force F-111 was shot down by surface-to-air missiles during the 1986 attack on Libya, while the fighter bomber was delivering a laser-guided smart munition. Brilliant robotic bombs may make such missions much safer for pilots and navigators. The U.S. Air Force plans for the Autonomous Guided Bomb (AGB) to enter its weapons inventory in early 1990. It will originally be used in a six-month high-speed test in which the AGB will be released from F-4 Phantoms. Future pilots of F-15, F-16, A-7A, and A-10 aircraft will also be outfitted with brilliant bombs.

ROTOR RPVs

The introduction of the helicopter not only revolutionized aviation but also altered the conduct of modern warfare. Being able to hover in a stationary spot, as well as perform vertical takeoffs and landings, led to the evolution of new missions, tactics, and capabilities for airborne systems. The use of helicopters transformed a large segment of the U.S. Army from a one-dimensional ground force into a quick-moving fighting body that could leapfrog over hostile forces and terrain. The navy was able to shift many of its antisubmarine and mine-sweeping functions from surface vessels to rotor vehicles. All three services adapted helicopters to search-and-rescue missions, and have used these versatile platforms to carry out special operations. As rotor craft have altered the shape of manned aviation, so are remotely piloted helicopters (RPHs) now revolutionizing the capabilities and mission profiles for unmanned aerial vehicles.

As mentioned in Chapter 2, rotary-wing drones were first experimented with in the 1960s, under the U.S. Navy's Drone Anti-Submarine Helicopter (DASH) program. This effort never gained much acceptance among the operational navy, due to the fact that these early tethered drones play havoc on other topside activities in their few demonstrations on surface vessels. Kaman Aerospace of Bloomfield, Connecticut, suffered a similar fate with their Ship-Tethered Aerial Platform (STAPL) effort of the 1970s, which was under contract to the U.S. Office of Naval Research. The STAPL, as did the DASH, had its advocates in the navy research and development community, but the many critics on the navy's operational side eventually prevailed, ending the program.

Today's improved technology has enabled rotor remotely piloted vehicles to be free-flown (without a tether). This, combined with current operational considerations, has made rotor RPVs attractive to the U.S. Navy, and to other military forces throughout the world. Although the U.S. Navy purchased the Israeli-developed fixed-wing Pioneer RPV in 1986 to meet its current needs, advanced-weapons researchers at the Naval Air Systems Command have had their eye on unmanned small-sized VTOL (vertical takeoff and landing) drones to perform future missions.

Rotor RPVs could operate from all types of ships, from the smallest patrol boat to the largest aircraft carrier, and provide over-the-horizon (OTH) coverage. They are capable of instantaneous omnidirectional movement in three dimensions and can maintain a fixed position relative to a specific ship or task force, thus matching surface-operations activity. Rotor RPVs can effectively carry out naval missions, such as ship-to-shore bombardment targeting and assessment, sonobuoy deployment, and use as a decoy in missions requiring a fixed position in order to simulate a hovering helicopter or stationary vessel. Remotely piloted helicopters (RPHs) require minimal training and skill for takeoff and recovery operations. They do not require the cumbersome equipment necessary for fixed-wing RPVs, such as launchers and recovery systems incorporating a parachute, net, or arresting cable.

The difficulties posed by launch and recovery operations at night, in adverse weather conditions, and on limited flight-deck space with unpredictable ship motions, which hampered the rotorplane RPV efforts of the 1960s and 1970s, have been overcome with today's technology. One drawback of rotor RPVs is that they are not easily air launchable from manned aircraft, a requirement that the U.S. Air Force considers essential for most missions, and that would also be useful to the navy.

The Sentinel

Canadair's CL-227 Sentinel, whose mode of operation is being emulated by some other RPH vehicle concepts, utilizes a haul-down system similar to that developed for the LAMPS helicopter used on surface combatants. A launch-and-recovery data link controls the CL-227 to within a few miles of the ship. At this point, the RPV's control is transferred from the operations room to the launch-and-recovery position, where the vehicle is maneuvered to a hovering position above the landing platform. A signal to the air vehicle releases a tether spool that carries a Kevlar line. The line is fed to a winch on the landing platform, which hauls down the rotor-plane RPV. Flight is accomplished through remote-controlled instructions in which an operator on board the vessel steers the RPH according to information received through the drone's sensors and via TV images picked up by the RPH's cameras. Some level of independent flight can be obtained through preprogrammed instruction, and future developments in artificial intelligence are aimed at giving such systems a completely autonomous capability.

The Sentinel (Figure 4.6) is especially attractive to naval analysts as a possible RPH candidate, or at least in serving as a test bed to demonstrate rotor drone concepts. Due to its unique hourglass configuration, this vehicle is known as the Peanut. The top of the vehicle houses a 50-hp gas-turbine engine. The center section contains an avionics module and attaches to the

Figure 4.6. CL-227 Sentinel with contrarotating propellers takes off from a launch vehicle. (Courtesy of Canadair.)

two contrarotating, three-blade propellers. The lower section includes the data link and payload, which can contain video, audio, infrared, and radar components, depending on mission profile. Launch weight for the CL-227 is 340 lb (127 kg) and it can lift a payload of some 60 lb (22 kg). Its height, including landing gear, is a mere 5.4 ft (1.6 m) and it has a body width of only 2.1 ft (0.6 m). The rotor disc diameter of 8.3 ft (2.5 m) allows the Sentinel to be operated from small vessels. It can stay in flight for up to four hours and has a maximum operating altitude of 9,200 ft (2,806 m) with a level speed of over 80 mph (129 kph). Because of the vehicle's low speed, some detractors view the Sentinel as an easy target. Several factors, however, make this RPH and other similar vehicles difficult targets to hit. Rotor downwash results in exhaust dilution, thus providing very little heat emissions for infrared-guided weaponry. The curvature of the CL-227's body and the lack of sharp angles lower the radar signature of the vehicle, providing a radar cross section of approximately 1 m^2. The rotor blades are made of low-reflecting composites, and radar-absorbing materials are used in selected areas. With no wings, the Sentinel presents a minimal visual target of less than 2-nmi (nautical miles) detectability. Its low rotor-tip speeds results in a very reduced noise level of less than 0.24-nmi detectability, making the vehicle ideal for certain covert surveillance missions. Perhaps the greatest advantage of the rotor RPV is that its cost is not much greater than many surface-to-air missiles, making it a very uneconomical target for an enemy to waste munitions on. The U.S. Army has also been evaluating the use of the Sentinel as a short- to medium-range operations surveillance system at the brigade level. This RPH is also being examined by the Canadians for use on their frigates and by Britain's Royal Navy as a possible decoy. The British experience in the Falklands revealed that chaff may divert a missile from a ship target only to have it lock on to another vessel. An RPH, however, could be sacrificed as a decoy to prevent a missile hit on a vessel. The Canadair CL-227 is currently in full-scale development and production is scheduled for 1988.

U.S. Rotor RPVs

The U.S. Army has been interested in the use of rotor RPVs as targets simulating Soviet manned helicopters. Hynes Aviation Industries, located in Frederick, Oklahoma, received an army contract to produce several of its H-5 RPHs, which resemble Soviet Hind helicopters (Figure 4.7). These vehicles have an overall length of 26.9 ft (8.2 m), a height of 8 ft (2.5 m), a cabin width of 54 in. (137 cm), and an empty weight of 1,700 lb (634 kg). The

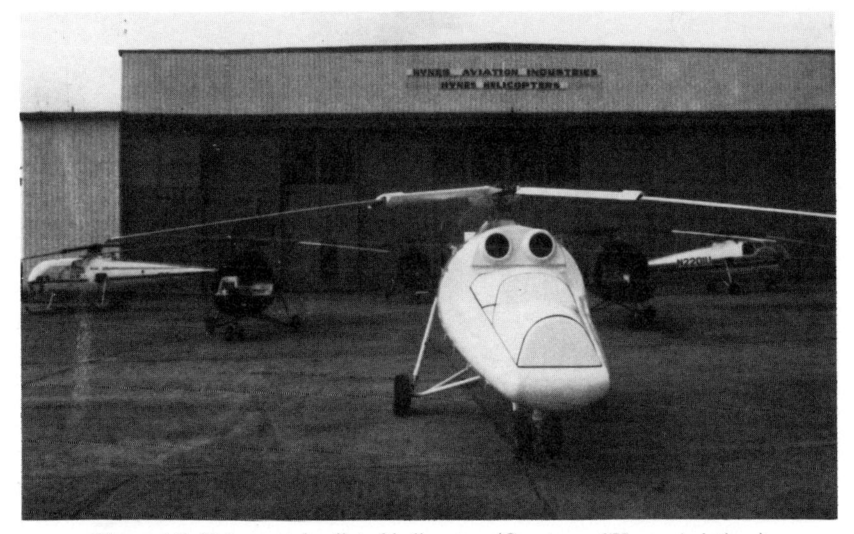

Figure 4.7. H-5 remotely piloted helicopter. (Courtesy of Hynes Aviation.)

army is using the H-5s as low-cost targets for radar-controlled antiaircraft guns and ground-missile tests and are considering buying some 400 RPH targets over the next few years. Hynes has proposed using its H-5 drone for high-risk "over-the-ridge" type observations, and as a platform for the remote pinpoint delivery of offensive weapons. In light of the direction towards the use of helicopters as air-to-air weapons platforms, munitions-carrying RPHs may be a viable alternative for certain missions in this high-threat environment. As indicated earlier in this chapter, the U.S. Army Office of Research was studying the use of "dogfighter" RPVs to obtain superiority above the forward edge of the battlefield. Perhaps rotorplane RPVs such as the H-5 would be viable candidates to perform such missions.

Teledyne Ryan developed a quarter-scale Mi-24 Hind D target RPH with a 50-hp engine (Figure 4.8). Although this vehicle was initially conceived of as an antiaircraft artillery target, Teledyne Ryan had been trying to convince the U.S. Marine Corps to support this rotor RPV technology as a baseline from which to develop a tactical surveillance version. Failure to do so has led the company to curtail further RPH developmental efforts. Rotor RPVs would be beneficial to the Marine Corps because the corps does not have the long-range requirements of other services, and they have little "beach-front" room for the launch/recovery and control vehicles needed by fixed-wing RPVs.

Aerodyne Systems Ltd. is involved in two rotor RPH developmental

Figure 4.8. One-quarter-scale Mi-24 Hind D target RPH. (Courtesy of Teledyne Ryan.)

programs. The Hornet is the drone version of Texas Helicopter's Wasp utility helicopter. The Pegasus CH-84 coaxial RPH externally resembles the DASH vehicle of the 1960s, although it is outfitted with the latest surveillance, reconnaissance, and communications technology. It is a fairly large vehicle with an empty weight of 745 lb (278 kg) and a maximum takeoff weight of 2,600 lb (969 kg). Rotor diameter is 20 ft (6.1 m) and the height is 9 ft (2.7 m). Forward-looking infrared (FLIR) systems and laser range finders/designators are being adapted for the Pegasus payload.

Aerotronics, Inc. of Miami is marketing two versions of their Dragonfly mini-RPH, which carries either T.V. camera or radar/infrared sensors. The RPH-1 is 8 ft (2.4 m) long, has a main rotor diameter of 10 ft (3 m), and its empty weight is only 70 lb (26 kg). It can carry 70 lb (26 kg) of payload, including fuel. It can reach speeds of 75 mph (122 kph) and has a maximum altitude of 10,000 ft (3,046 m). The larger RPH-2 is 9 ft (2.7 m) long with a rotor diameter of 12 ft (3.6 m). Its empty weight is 80 lb (38 kg) and it can carry a 70-lb (26-kg) payload and fuel. The Dragonfly is controlled by high-frequency proportional radio and has been used by Martin-Marietta Corporation to take video pictures of the Air Defense Anti-Tank System (ADATS) during its testing. The Dragonfly's potential military missions include reconnaissance, surveillance, attack, transport of small items, decoy, relay, and aerial photography.

European and Chinese Rotor Research

The French Dorand D5.7 was an attempt in the early 1970s to use a remotely controlled gyro-glider to softly land loads dropped from aircraft. The system never went beyond some experimental test flights. In the mid-1970s, the French and West Germans initiated a joint program called the Argus to develop a tethered autonomous-radar battlefield-surveillance system. It used the Dornier-built Kiebitz rotor-borne platform with French LCT Orphee reconnaissance radar. The vehicle was successfully demonstrated on several occasions and in May 1979, it reached a height of just under 1,000 ft (305 m), and transmitted radar data to the ground station via cable. Boeing Corporation pursued applications for the Kiebitz rotor platform in the United States in return for Dornier's advocacy of Boeing's Compass Cope RPV in West Germany. The Argus system nonetheless was eventually cancelled in favor of the Argus 2, which is a free-flying RPH. A model of the Argus 2 surveillance system, which utilizes the MTC II RPH, was unveiled by Dornier at the 1985 Paris Air Show. The vehicle is equipped with a 40-hp piston engine and has coaxial counterrotating rotors. It will carry a maximum payload of 132 lb (49 kg) and has an endurance of two hours. Potential missions include battlefield surveillance, reconnaissance of ground targets, and detection of helicopter targets.

The British ML Aviation SPRITE mini-RPH (Figure 4.9) has been test-flown at the Pentagon and at Dulles Airport outside of Washington, DC. This small vehicle has a body diameter of only 2 ft, 1 in. (650 mm), a height of 2 ft, 9 in. (900 mm), and a rotor diameter of 5 ft, 3 in. (1,600 mm). It can carry a sensor package weighing 13.2 lb (6 kg), can reach a maximum speed of 62 mph (110 kph), and has an endurance of 2.5 hours. It is designed for tactical reconnaissance, artillery observation, NBC "sniffing" detection, target marking, and security surveillance. The U.S. Air Force is considering SPRITE as an air base damage-assessment vehicle. The U.S. Navy is interested in the SPRITE as a system to check on floating objects before approaching them with ships. It appeals to the U.S. Army as a possible

Figure 4.9. The tiny SPRITE mini-RPH. (Courtesy of ML Aviation.)

forward battle-tank designator, and to elements of the U.S. Rapid Deployment Forces, because this RPH and its support equipment can be carried in a land rover.

British sensor manufacturer Vintem Ltd. got into the rotor-RPH airframe business as a result of not being able to find a suitable platform for some of their new sensors. They decided to build their own unmanned Autogyro vehicle. The Autogyro resembles a conventional rotary-wing helicopter, except that the rotor blades are unpowered and remain at zero incidence. The horizontal thrust comes directly from the engine and lift is generated by inclining the rotor away from the relative airflow. Although not capable of vertical takeoffs and landings, its performance fits into the ranks of short takeoff and landing (STOL) aircraft.

The Chinese Nanjing Research Institute has developed a small experimental rotor RPH named the Nanjing 2-2. It has a single main tail-rotor configuration powered by a 5-hp engine. The results of 2-2 tests are being used to develop more-advanced RPHs.

Nonrotor VTOL Aircraft

Some nonrotor, yet VTOL-capable, RPVs are being developed for military missions (Figure 4.10). AAI Corporation, headquartered in Baltimore, is working on a high-speed, high-altitude VTOL RPV for real-time T.V. reconnaissance. Dornier has been involved in a very innovative vectored-thrust VTOL RPV. This research was originally begun following World War II in the U.S. by A. Lippisch, who was the German engineer responsible for development of the Nazi-era Komet rocket-powered interceptor. Dornier has since carried on Lippisch's research through contracts from the West German Federal Ministry of Defense. The turboshaft engine obtains vertical lift by the deflected thrust from a shroud via a deflection chamber and nonmoving cascade vanes. This vehicle is capable of achieving high subsonic speeds. Dornier has teamed up with the British HSD Ltd. to develop a maritime version of this vectored-thrust VTOL, named the Maritime Aerodyne. The concept involves one turboshaft engine driving two nine-blade shrouded-fan propellers with cascade-type thrust-vector control vanes in the rear of each duct.

The U.S. Marine Corps, with assistance from Sandia National Laboratories, has been working on an 80-lb (36-kg) shrouded-fan flying platform called the Airborne Remotely Operated Device (AROD), which has a cross-section diameter of only 3 ft (0.9 m). An electrically powered prototype has already been demonstrated, and a more advanced 25-hp gasoline- or diesel-engined system is in development. The AROD is designed to be used by marines at the battalion and company level in direct support of the front-line commander. It will allow the commander to peek

Figure 4.10. Vectored-thrust VTOL RPV. (Courtesy of Dornier.)

around natural obstacles such as hills, out to a distance of 15.5 mi (25 km). Besides providing reconnaissance, the AROD, with its navigation and laser range finder, can also serve as an effective artillery forward observer. Potential missions also include minefield detection and neutralization, NBC agent monitoring, and antiarmor activities. The AROD can be a tank killer by mounting hypervelocity rockets on it, such as the small SPIKE rocket developed for the U.S. Army. It would be firing them from above, where the tank is most vulnerable. The AROD relies on remote presence in which a fiber optic cable transmits images from a camera onboard the vehicle to a heads-up display on the face plate of the operator's helmet. The operator relies on joystick control to maneuver the AROD. A sophisticated gyroscopic system enables the platform to remain in a stable hover in wind speeds up to 30 mph.

Summary

The future battlefield is likely to contain strange looking aircraft, capable of instantaneously darting and bobbing back and forth and up and down, and resembling the movement of a hummingbird in flight. Remotely piloted helicopters and other VTOL RPVs will have a profound impact on the development of unmanned vehicles and their tractical uses. The capabilities posed by rotor drones can have as dramatic an influence on contemporary warfare as did the introduction of manned helicopters.

LIGHTER-THAN-AIR UNMANNED VEHICLES

The need to perform aerial surveillance and early warning missions economically and over long durations of time has led to the "rebirth" of lighter-than-air vehicles (LTAs), otherwise known as airships. The U.S. Navy and other nations' fleets are examining and testing new airship configurations and concepts designed to provide over-the-horizon detection of low-level cruise missiles. Airships that incorporate new technology and designs can outperform aircraft systems in certain mission applications. Unmanned airships, including aerostats and remotely controlled drone "blimps," can provide some unique capabilities to perform surveillance and other missions more effectively and efficiently than manned versions.

Aerostats

Ground-based radar systems have severe limitations in detecting low-flying aircraft and can be obstructed by natural terrain. The cost of providing around-the-clock radar coverage by aircraft has become prohibitive in certain areas. Aircraft also have certain weight and configuration restrictions limiting the size of the radar. Tethered aerostats, which are blimp-shaped observation balloons tied to the ground by strong cables (Figure 4.11) have helped to overcome some of these weaknesses, and have played important law enforcement and military roles since the 1960s. Hovering at altitudes up to 15,000 ft (4,570 m), these vehicles have sophisticated radar and sensor devices. The Seek-Sky-Hook program of the U.S. Air Force relies on aerostats to provide radar coverage over the Florida Straits to guard against aircraft incursions from Cuba. The U.S. Customs Service uses aerostats to watch over air traffic and to detect and intercept suspected drug smugglers. In the Caribbean, the U.S. Coast Guard is flying an aerostat from an offshore vessel that is used to relay information to patrolling vessels. Westinghouse Electric Corporation's TCOM subsidiary and ILC-Dover Corporation are major manufacturers of sophisticated aerostat systems ranging from 60 ft^3 (15.3 m^3) to very large airships of over 250,000 ft^3 (63,649.7 m^3).

Other LTA Concepts

Several innovative remotely piloted airship concepts have been under development. Developmental Sciences, Inc. of California has successfully test-flown their Remotely Piloted Mini Blimp (RPMB), which is especially suited for urban surveillance. This prototype unmanned airship is 16 ft (4.88 m) long, has a diameter of 4 ft, 1 in. (1.25 m), and can carry a 3.5-lb (1.3-kg) payload as well as an electronics package, movie camera, and propulsion system. A larger version of approximately 50 ft (15.25 m) has been proposed for law enforcement officials and the U.S. Federal Aviation Administration (FAA) to perform such duties as traffic monitoring and other surveillance missions. The low noise level of an RPMB-type vehicle could make it an effective means to perform counterterrorist missions by discreetly surveilling urban hide-outs. The Japan Experimental Aircraft Association has built a small 25-ft (7.62-m) pilotless research airship named the "Flying Submarine" that incorporates a unique wing fitted beneath the gondola to provide additional lift and vertical control.

The U.S. Navy gave funding to the Lockheed Missiles and Space Company to produce a conceptual design for a high-altitude airship named the Hi-Spot. This vehicle would be capable of hovering in one location at an altitude of 70,000 ft (21,350 m) for periods up to 100 days. Its 500-ft (152.5-m) nonrigid design would carry 5 million ft^3 (1,272,994 m^3) of helium. Intended Hi-Spot missions include air/sea surveillance, communications relay, and sensor readout. This type of LTA could observe military operations over large stretches of land and water, and would be effective in areas not covered by satellite surveillance. The Royal Navy could have surely used one during the Falklands war. The U.S. Navy has also funded Martin-

Figure 4.11. Aerostat moored to ship. (Courtesy of TCOM.)

Marietta Corporation to develop their High Altitude Superpressure Powered Aerostat (HASPA) concept. This remotely piloted aircraft would be 333 ft (101.5 m) long and have a diameter of 67 ft (20.4 m), and could reach altitudes up to 68,000 ft (20,740 m).

The Soviets have also experimented with remotely piloted LTAs, and have flight-tested a model designated the Angren 84. This vehicle is powered by two engines, which can be rotated, and has a length of 30 ft (9.1 m) and a diameter of 10 ft (3 m). The Angren 84 will likely serve as a prototype for a much larger LTA designed to perform cargo hauling and aerial photography.

Some LTA manufacturers have developed remote-controlled airships to serve as prototypes for larger manned and unmanned systems. The Magnus Aerospace Corporation has produced three innovative remotely piloted drones that are spherical-shaped, rather than the typical cigar-shaped airship. This vehicle, known as the LTA 20-1 (Figure 4.12), is in fact a hybrid combining the buoyant lift of helium, the dynamic lift of engine thrust, and the aerodynamic lift of the Magnus effect. The Magnus effect is created during forward flight when the spherical envelope rotates around its horizontal axis. The top of the sphere rotates away from the direction of travel, creating a velocity differential between the top and bottom. A pressure differential is caused, which results in additional aerodynamic lift. Three six-channel radio transmitters have been used to remotely test-fly the Magnus LTA 20-1 in an aircraft hangar.

A deltoid-shaped rigid airship named the Aereon 26 has been test-flown by Aereon Corporation of Princeton, New Jersey. This hybrid design adds helium static lift to aerodynamic wing-lift. The U.S. Strategic Defense Initiative (SDI) Program Office has examined the use of airships to serve as coastal early warning systems to guard against submarine-launched ballistic missiles (SLBMs). An unmanned deltoid-shaped airship cruising at an altitude of 40,000 ft (12,200 m) could make an effective laser-weapons platform to shoot down incoming ICBMs. The airship would be above atmospheric conditions that could interfere with the beam's propagation, and much of the laser's piping and circuitry could be maintained internally within the airship. Being unmanned, it could spend weeks on-station without having to change crews or be refueled. The deltoid shape would also have inherent stealth advantages over cigar-shaped LTAs.

Stewkie Aerodynamics, a small firm based in Dorset, England, has produced some interesting remote-controlled inflatable "lifting bodies" that have undergone trials with the Royal Navy. A 19-ft (5.8-m)-long vehicle with a wingspan of 22.8 ft (7 m) is called Float-a-plane. Its two dc electric motors provide silent propulsion, and combined with its transparent inflatable material, it could carry out surveillance work with little chance of being detected by radar or visually sighted. The British Ministry of Defense

Figure 4.12. LTA 20-1 spherical airship being remotely maneuvered in hangar test flight. (Courtesy of Magnus Aerospace.)

(MOD) has expressed interest in a more outlandish Stewkie design called the Sauceror. This LTA RPV consists of two inflated ribbed spheres that are contrarotating. The vehicle would have rudder and elevation controls in the form of winglets at the periphery of the discs on which motors would be attached. A computer would be on-board to provide vertical and directional control, and theoretically the Sauceror would be extremely stable and could carry heavy loads. ILC Dover has test-flown an inflatable 5.06-ft (1.55-m)-wingspan RPV and has proposed a 75.21-ft (23-m)-wingspan RPV that could carry a 200-lb (91-kg) surveillance payload.

The development of innovative LTA designs using new technologies has led to the reintroduction of airships into the armed forces structures of several nations. The advantage of using unmanned versions to perform airship missions should be taken into account by key weapons decision makers.

KITES AND PARAFOILS

The U.S. Marine Corps has been examining the use of a lightweight, tethered airborne-surveillance system to provide an elevated look over the "next ridge" to an amphibious force just landing on the beach. Such beachhead positions often lack adequate surveillance and target-acquisition information. The tethered surveillance robot, which might resemble a very sophisticated kite, is one possible approach to achieving this objective.

Lockheed Missiles and Space Company of Sunnyvale, California, has been designing and testing all-fabric unmanned rectangular parachutes called parafoils. These electronically guided vehicles are designed to carry air-dropped cargo to pinpoint-accurate landings. The parafoils have been powered and unpowered, and are radio controlled by a ground operator. They can also be directed from aircraft. Besides resupplying troops from standoff aircraft, the robotic parafoils can also perform reconnaissance work. The U.S. Army's Research, Development, and Engineering Center in Natick, Massachusetts, has also been developing an airdrop system designed to glide from high altitudes to a pinpoint target area. The system allows for the accurate clandestine delivery of supplies for special operations. Using it allows an aircraft to launch the ram-air gliding parachute at a safe distance away from the recovery area. An airborne homing and guidance system and a transmitter controller guides the system to a designated area, which is followed by the flaring of the canopy for a soft landing.

TEST-FLIGHT PROTOTYPES

Using scale-model drones of manned aircraft to test vehicle designs, advanced composite and metallic structures, and flight and propulsion control-system designs can save the lives of test pilots as well as a great deal in research funding. High-risk technology can be tested sooner and at lower cost by using RPVs, especially because the aircraft do not need to be man-rated, which is an expensive and time-consuming process.

NASA's Test-Flight RPVs

NASA's Dryden Flight Research Center first applied a technique used for flying RPVs to a three-eighths scale model of the F-15 Eagle fighter. The replica was launched from a B-52 and flown by a pilot located in a ground cockpit that was complete with flight controls and a T.V. screen. Commands were fed through a computer and then telemetered to the RPV. The F-15 replica RPV was extremely useful in testing the aircraft's configuration in dangerous spin tests and in other high-risk situations, and when cost precluded building a full-scale manned aircraft.

NASA developed an oblique-wing RPV that allowed the wing to be rotated on its center pivot so that it could be set at its most efficient angle for the speed at which it was flying. The aircraft could thus take full advantage of the swept-wing configuration at high speeds while offsetting some of the normal drag-producing air disturbances. This research was very useful to LTV Corporation's advanced design, which mates the fuselage of an A-7 aircraft with an oblique wing.

The NASA DAST (Drones for Aeronautical and Structural Testing) program involved replacing the standard wing of the Teledyne Ryan BOM-34F Firebee II target drone with a supercritical airfoil wing featuring an active flutter-suppression system.

The HiMAT

The prototype-testing RPV that received a great deal of publicity was the HiMAT aircraft built by Rockwell (Figure 4.13). The HiMAT program was designed to test advanced designs as to their transonic maneuverability. The subscale-model RPV bridged the gap by furnishing data that wind tunnel testing, simulators, and ground tests failed to provide. It offered the next best thing to actual full-scale manned flight testing. The HiMAT vehicle was launched from a B-52 at an altitude of 45,000 ft (13,696 m). It was flown in a similar fashion to that of the F-15 scale-model RPV, by a test pilot from the ground facility. The ground cockpit contained a throttle, stick, rudder pedals, and sensor displays. A computer converted the pilot's actions into electronic telemetered signals to the HiMAT vehicle. The RPV's onboard computer sent signals through the digital fly-by-wire system to the flight-control surfaces. The HiMAT telemetered thousands of bits per second of real-time flight data back to ground computers. Its skid-type gear enabled the vehicle to land on dry desert lakebeds. The two HiMAT RPVs manufactured by Rockwell were used to demonstrate high-G maneuvers in the transonic flight regime of 600–800 mph (965–1,287 kph). The HiMAT was about half the size of the future manned fighter vehicle concept it resembled. The RPV was 22.5 ft (6.9 m) in length, had a wingspan of 15.6 ft (4.7 m), and a height of 4.3 ft (1.3 m). It was powered by a General Electric J85-21 afterburning engine and had a launch weight of 3,400 lb (1,544 kg).

Other aircraft technologies, including high-aspect-ratio fuel-efficient wings, forward-swept wings, skewed wings, and low-observable geometric designs, can be tested on RPVs.

Figure 4.13. HiMAT prototype test RPV. (Courtesy of Rockwell.)

SUMMARY

In the realm of air-breathing vehicles, it is clear that numerous robotics technologies are being explored, although with limited procurements. Some of the most advanced technologies and materials are being joined with sophisticated sensors, computers, and command and control equipment. Any introduction of these systems into the military inventory will be determined by performance and by the willingness of the air force leadership to accept these unmanned systems into its plans. This same process is proceeding within the U.S. Navy as robotic technologies become more important to surface and underwater warfare, as we will see in the next chapter.

NOTES

1. Arnold's speech was retold by General Lawrence A. Skantze, commander, Air Force Systems Command at the Air Force Institute Association of Graduates 4th Biennial Technical Symposium, during the commandant's breakfast, held at Wright-Patterson Air Force Base, Ohio, on 11 October 1985.
2. In the U.S. Air Force's 1987 budget report to Congress, Russell A. Rourke, former air force secretary, and General Charles A. Gabriel, air force chief of staff, delineated what they saw as the positive and negative features of RPVs. The primary advantages were the lower RPV cost when compared to manned systems, and that personnel are not exposed to the threat. Disadvantages included lack of flexibility on the part of RPVs, because the targets are preprogrammed; an inability to react to or avoid mobile air or ground threats; system susceptibility to inertial navigation errors; possibly greater difficulties in deployment than exists for manned systems; and system constraints in peacetime applications and training. They concluded that unmanned air reconnaissance vehicles (UARVs) are needed to complement, not replace, manned systems.

5 Current Operational Use of and Developments in Unmanned Naval Vessels

Divers and submariners face an environment that is naturally inhospitable. When combined with warfare conditions, it can make such underwater missions extremely hazardous. In order to reduce the danger, certain manned missions have been replaced or supplemented with the use of robotic undersea vehicles. These unmanned submersibles, commonly referred to as Remotely Operated Vehicles (ROVs), can take some of the workload off of highly skilled personnel by performing the missions that could entrap or disable manned vehicles or divers. If the ROV is destroyed, only machinery, not lives, are lost. By having the option of using robots, the mission commander can be more selective and reserve the higher-priority missions, or those of particular suitability, for manned systems. Budget constraints and skilled-manpower shortages have also facilitated the use of ROVs. Such systems are cheaper and smaller than manned submersibles because they do not have to incorporate expensive life-support features. Without the need for physiological-related procedures, ROV operational usage can also be more efficient and timely.

ROVs are currently being used in a number of naval missions, including the mapping of the ocean floor's topography, inspection of sunken ships and aircraft, retrieval of ordnance and objects, rescue of manned submersibles, installation of antisubmarine-warfare listening devices, and for the detection and neutralization of enemy mines. Conceptual autonomous submersibles and submarines are being designed for offensive attacks on enemy ships, submarines, and harbors. Unmanned boats, frequently used for target practice, are making a comeback (from the World War I era) as reconnaissance and attack craft.

TIED-DOWN ROVs

Most remotely operated vehicles in operation today utilize an electromechanical umbilical cord connected to a surface vessel. This tethered line's advantages include the following: (a) provision of an unlimited energy supply, (b) capacity for high data-transmission rate, and (c) a safety link to the support ship. However, the cable restricts maneuverability and accessibility, and often requires extensive maintenance and replacement.

U.S. Navy Tied-down ROVs

The U.S. Navy has an Unmanned Vehicles Detachment operating out of the Submarine Rescue Unit at NAS North Island, California (near San Diego). This group of elite people and robots conducts search, location, rescue, and recovery operations throughout the world. The majority of this detachment's ROVs have been of the tethered type, such as the RCV-150 built by Honeywell Corporation's Hydro Products Division. The navy leased this vehicle to the Navy Ocean Systems Center (NOSC) laboratory in Hawaii to collect operational and logistic support data for a study. The RCV-150 was used to search and recover various objects. Its abilities were clearly demonstrated in a mission when the vehicle's arm was able to hook into an anchor's chain in 360 ft (110 m) of water, allowing the anchor to be hauled to the surface.

CURV III and Deep Drone

The navy's ROV workhorse has been the Cable Controlled Underwater Recovery Vehicle (CURV III), which has accomplished everything from fixing acoustic equipment and rescuing trapped submariners to recovering torpedoes. Beginning in 1984, CURV III usage began to be phased out with the introduction of the Deep Drone, a vehicle that can perform emergency rescue operations and conduct deep-ocean salvage operations. Deep Drone can reach depths of 6,000 ft (1,828 m), and travel at speeds of 3 knots. It is outfitted with one black/white and one color camera, a 360-degree search sonar, and a bottom-navigation system that incorporates both a seven- and a four-function manipulator. Deep Drone is being managed by the Naval Sea Systems Command, which has insured that the ROV has continuously been updated with the latest robotics technology advances. Several accomplishments of the CURV III and Deep Drone ROVs were described in Chapter 2.

Recon 4, Sprint, and Scorpio

An example of a major recovery effort, undertaken in part by unmanned submersibles, was the search following the tragic January 28, 1986, *Challenger* space shuttle explosion. To better understand its cause, investigators had to put together the pieces of the wreckage, which resembled an effort of paleontologists reconstructing a dinosaur. Robot submersibles played a vital part in sifting through the fossilized remains of the myriad failed rockets launched over Cape Canaveral during the last three decades, and shipwrecks dating back to the time of the *conquistadores*, to find the *Challenger* debris.

During the first few days, the navy, in conjunction with the coast guard and air force, sent numerous ships, patrol aircraft, and helicopters to sweep across the 6,000-mi^2 (15,540-km^2) of ocean surface, gathering evidence. The search soon shifted to the ocean floor. Surface vessels had obtained some sonar contacts suggesting possible shuttle pieces, but bad weather and extreme depths, sometimes exceeding 1,000 ft (305 m) where much of the debris was located, prevented standard diving operations. Salvage of certain pieces, such as the satellite rocket, was considered extremely dangerous to divers and manned vehicles because its fuel could be ignited underwater by a stray spark.

Three robot submersibles were deployed to the scene. The very small Recon 4, and the 240-lb (109-kg) Sprint unmanned submersible built by Perry Offshore Company of Riviera Beach, Florida, provided camera coverage to determine what the wreckage "pinged" by the sonar actually was. They provided the first glimpse of part of the *Challenger*'s crew cabin. A larger 2,500-lb (1,135-kg) remotely operated vehicle named the Scorpio, which is in the same class as the CURV III and Deep Drone vehicles, used its two manipulators to actually grab some of the shuttle pieces and return them to the surface. One manipulator is a five-function arm controlled by a joystick resembling one used in an arcade game. The other seven-function arm is controlled by a "master–slave" system. This system is composed of a smaller replica arm called the master, located in the support ship's control room. When the master arm is moved to the desired position, a computer in turn measures the new coordinates and transmits the corresponding control signals through a cable to Scorpio's slave arm. The manipulators are electrohydraulic, requiring up to 27 gal (102 L) of hydraulic fluid per minute.

The robotic submersibles helped "pave" the way for follow-up searches by identifying the best objects for additional inquiry. Manned vehicles, such as the Johnson Sea Link II research submersible and the navy's NR-1 nuclear-powered deepwater research and engineering submarine, joined the search effort several weeks later and provided more in-depth investigation in the areas pioneered by the ROVs.

Views of the Titanic from a Minisubmersible

The U.S. Navy has sponsored development by the Woods Hole Oceanographic Institution of an unmanned 28-in. (71-cm) minisubmersible called Jason, Jr., which houses a video camera capable of surveying 170 degrees of the visual field. This vehicle, which the media have labeled the "robot camera," is launched from the bow of a manned submersible or submarine and is guided by instructions transmitted through a 200-ft (61-m) cable tethered to the minisubmersible. In order to demonstrate the capabilities of Jason, Jr., the Woods Hole Oceanographic Institution operated the minisub's camera from the manned submersible Alvin, enabling it to take some spectacular photographs of the interior of the luxury liner *Titanic*, which was lying beneath 12,500 ft (3,809 m) of water.

U.S. Tethered ROVs and Mine Destruction

Over the past ten years, tethered ROVs have been introduced into the tactical arena as mine-destruction vehicles. The U.S. Naval Sea Systems Command awarded a $30.1 million contract in September 1984 to Honeywell's Marine Systems Division for the production of an unspecified number of Mine-Neutralization Systems (MNSs). The MNS uses a 2,500-lb (1,135-kg) ROV to find, classify, and ncutralize underwater bottom and moored mines (Figure 5.1). Honeywell began development of the MNS in

Figure 5.1. Mine Neutralization System being lowered into water from mine-sweeping vessel. (Courtesy of Honeywell.)

1978, and has conducted testing at the Naval Ocean Systems Center and the Naval Coastal Systems Center. Honeywell produced the MNS shipboard control console, performance monitors, sonar, television displays, and vehicle status information monitor. The ROV itself is manufactured by Hydro Products, a division of Honeywell's wholly owned subsidiary, Tetra Tech. Hydro Products is a well-known manufacturer of commercial submersibles, such as the RCV-125, 150, and 225 models, which have been leased by the navy to perform salvage and object-recovery tasks. The navy will deploy the MNS system on its next class of minesweepers, and possibly on its Arapaho-type merchant vessels, which are reconfigured into mine-clearing ships in wartime by using containers.

Honeywell's MNS furnishes power to the ROV through a neutrally bouyant umbilical cable, which gives it unlimited endurance. The handling system can reel out 3,500 ft (1,067 m) of cable at a speed that keeps up with the ROV's 6 knots. In the typical mission, initial target-detection and vehicle-guidance information would be provided by the ship's sonar. An acoustic-position indicator would track the ROV's progress and be used to steer it in the correct direction. Midway, the ROV's own high-resolution sonar would be switched on and take over guidance input from the ship sonar. As the robot submersible arrives on the scene of the mine, a low-light, pan/tilt T.V. furnishes an image allowing the operator on-board the ship to precisely maneuver the vehicle. Upon command, the ROV neutralizes the mine with cable cutters or a dropped charge.

European ROVs and Mine Removal

France's PAP-104

The Société ECA of France marketed one of the first unmanned submersibles specifically dedicated to mine removal. Since the mid-1970s over 250 units of this system, known as the PAP-104 Mine Disposal System, have been sold to six European navies and Australia. The 1,540-lb (700-kg) vehicle has a 8.9-ft (2.7-m) cylindrically shaped watertight body. On each side is a thruster, on the top are a bobbin, transducer, and flashing light and mounted on the bottom are a searchlight, guide rope, and the charge. The Mark III version of the PAP-104 proved quite effective at clearing mines for the British during the Falkland Island conflict. The Mark IV can reach depths of 981 ft (300 m). The latest variant, the Mark V, has new electronics and teletransmission.

Inside the PAP-104 is a T.V. camera that views through a window in the forward section. The "guts" of the ROV houses an electronics cabinet, gyroscope, and power supplies. A battery is positioned in the rear.

In mine-hunting operations, the ROV is fitted with a charge to destroy or neutralize the mine. Once the charge has been dropped near the mine, the mine-hunter ship and ROV move to a safe distance away. Between fifteen and thirty minutes after the charge is released, a grenade is thrown that provokes the charge's explosion. The PAP-104 cuts moored mines by having the operator maneuver the ROV to the mine rope and then placing it between the two arms of the cutter with the help of the T.V. camera. The cutter is released and the PAP-104 is moved back to the ship. Twenty minutes later, a delay-firing unit triggers the explosion and the mine rope is cut. The mine floats up and is destroyed on the surface.

Italy's MIN

The Italian MIN consortium, made up of the two partner companies, Ellettronica San Giorgio-Elsay and Riva Calzoni, has built four of their MIN (Mine Identification and Neutralization) ROVs for the Italian navy for use on board the new Leric-class mine hunters. The MIN is a small 264-lb (120-kg) wire-guided underwater vehicle that receives instructions from an operator sitting at a console on the support ship. The console has a joystick to adjust speed and direction and controls for charge activation, ballast release, presetting the cutting device, and on/off functions for the T.V. and sonar. Turnaround time involved in preparing the MIN for a new sortie takes just twenty minutes. The Italian ROV's coaxial cable length is 3,280 ft (1,000 m), but its actual operational range from the minehunter ship is more like 920 ft (250 m). It has a maximum operational depth of 492 ft (150 m), and a maximum speed of 5 knots.

In order to hunt for bottom mines, the MIN is outfitted with an explosive charge and then lowered into the sea. The ROV slowly sinks as the umbilical cable begins to unwind from a reel in the vehicle. An auxiliary console operator on the stern of the minehunter ship steers the ROV away from the ship. When it is at a safe distance, control of the MIN is turned over to the operator at the main console. When the ROV reaches a depth of 33 ft (10 m), as displayed on the main console, the operator steers the ROV into the beam of the minehunter's sonar. By having the information on the position of the unknown object relative to the ROV displayed on the minehunter sonar console, the operator can then steer the ROV towards the target. This is accomplished by orientating the main propeller shaft by means of the joystick. An echo sounder indicator on the main console provides the distance between the bottom and the vehicle. As the ROV gets within 6.6 ft (2 m) of the mine, the operator reduces the forward speed by adjusting the main propulsion controls and bringing into use the forward and after thrusters, which provide precise movement. The ROV's T.V. camera and illuminating lamp aid the operator in identifying the object as a mine. The vehicle's sonar can be used to locate and identify mines in bad weather. If the object is a mine, the operator maneuvers the ROV into a position to drop a charge adjacent to it. When the MIN drops the charge, the ROV becomes instantly buoyant and starts to rise until it has reached a safe distance above the mine; it is then maneuvered back to the ship. This whole operation usually takes a well-trained crew about twenty minutes. The MIN uses an explosive cutter mechanism similar to Honeywell's MNS system to neutralize moored mines.

British–Canadian Trail Blazer

The British firm Fairey Hydraulics Ltd., in association with International Submarine Ltd. of Vancouver, Canada, has developed the Trail Blazer series of tethered-mine countermeasures ROVs. These ROVs range in weight from 546-lb (248-kg) to 1,698-lb (772-kg) and have speeds varying from 2.5 to 5.8 knots. They can all reach depths of 1,640 ft (500 m). The Trail Blazer's propulsion system consists of five hydraulic thrusters: (a) Two mounted on the aft end furnish forward and backward motion; (b) two amidship thrusters provide lateral motion, and (c) the remaining thruster, also located amidship, provides vertical motion. A control console joystick directs the longitudinal and lateral thrusters. Moving the stick away from the operator provides power in the forward direction; moving the stick towards the operator propels the ROV in the astern direction. Rotation of the stick provides differential pressure to the port and starboard thrusters, allowing the vehicle to be steered. Sideways movement of the stick results in lateral motion. A separate control on the console operates the vertical thruster. The Trail Blazer's mission profile is similar to that of other tethered-mine countermeasures ROVs, except that its high payload capacity allows it to transport a number of mine disposal charges. Instead of returning to the ship after setting each charge, it can place a number of charges during each run.

Other Mine-removal ROVs

West German firm MBB has developed the 11.5-ft (3.5-m), 2,970-lb (1,350-kg) PINGUIN B 3 ROV. Although remote controlled, the PINGUIN incorporates more automatic movement through programmed instructions than do most of the other ROVs. This is meant to reduce operator stress. The PINGUIN's mine-hunting mission sequence begins when the ship's sonar system sweeps several hundred feet ahead. When an object with "minelike" features is detected, the PINGUIN is launched and then guided visually on the surface through remote control, which steers it in the direction of the sonar beam's relative bearing. At a certain point, the target approach is switched to the automatic mode, and the ROV follows a preselected course, speed, and diving depth. When the ROV is in vicinity of the mine, remote control is resumed through the operator's use of a T.V. camera mounted on the PINGUIN. The PINGUIN carries two charges, so that after dropping one near the first mine it can go after a second object in the sonar's range. When the target-approach operations are completed, the ROV is guided back to the surface and then to the ship.

The Swedish firm SUTEC has introduced their mine-destruction ROV, Sea Eagle and has sold a number to the Swedish navy. The commercial Sea Owl ROV, from which the Sea Eagle is modified, was used by the British in 1984 to help clear mines in the Red Sea and is credited with discovering a Soviet mine never seen before.

This successful adaptation of a commercial ROV to perform mine countermeasures has led to some speculation on how such ROVs could be used to supplement military systems in time of warfare. This topic was addressed at the International Symposium on Mine Warfare Vessels and Systems held in London, in June 1984 in a paper presented by William H. Key, an executive with Klein Associates of Salem, New Hampshire. Key related how commercial offshore equipment, including remote-controlled submersibles, by using the right tactics could provide significant mine countermeasures at minimal cost. He described the DART, manufactured by International Submarine Engineering Ltd., as a lightweight commercial ROV that could be adapted to a military role. Although this may present some difficulty in meeting some military performance specifications, the alteration of operating tactics to permit variation from the standard specifications could allow the use of such vehicles without a serious degradation of performance. The DART would not meet certain military requirements such as those for magnetic field, mechanical noise, and acoustic noise emissions, which could result in a significant risk of destruction from the inadvertent detonation of mines. The cost of these commercial ROVs, however, is quite low, approximating the cost of a mine, and are thus considered expendable. Therefore, the intelligent use of commercial robotic submersibles could help offset the inadequate number of military ROVs and the overall lack of antimine warfare preparedness in wartime.

REMOTE-CONTROLLED FREE SWIMMERS

Tethered ROVs are limited in their maneuverability and speed. They also have to be in close proximity to the control ship. There has been an effort to develop tether-free swimming unmanned submersibles. One design thrust is to develop techniques for transmitting information and control signals at a high data rate between the support ship and ROV.

Acoustic Communications

Current systems that use acoustic communications have a number of serious drawbacks.

1. The greater the distance between the ROV and the ship transmitting the acoustic signals, the more quickly the sonic energy dissipates. The acoustic power output must therefore grow geometrically as the robot submersible moves away from the surface ship.

2. Acoustic communications allow for only a very slow rate of information transfer. Only 10,000 bits per second can be transferred under water, which is a tiny fraction of the 66 million bits per second needed to transmit a conventional T.V. picture. Current acoustic ROVs use slow-scan video systems that send a few frames at a time to the operator, rather than a continuous moving picture. Most of the data are recorded on a videocassette that can be viewed later, after the ROV rendezvous with the ship. The slow-scan images serve mainly to reassure the operator that the system is still functioning properly.
3. Acoustic signals can also experience sonic interference from things such as the sea floor, changes in water temperature, and large objects, which can severely degrade the signal.
4. Because acoustic signals rely on the speed of sound, approximately 5,000 ft (1,524 m) per second, they travel much slower than do electromagnetic signals going through a cable, which move at close to the speed of light. In operating at a depth of 1.5 mi (2.4 km) it would take over three seconds for an acoustic signal to reach the ROV from the ship. This complicates remote-control operations and requires the operator to extrapolate the ROV's probable path between transmissions.

Most of the military acoustic ROVs are research vehicles used to gather oceanographic or engineering data. The Applied Physics Laboratory (APL) of the University of Washington in Seattle has been working with the SPURV (Self-Propelled Underwater Research Vehicle) since 1963. The majority of its runs have been conducted from U.S. Navy AGOR-type ships. It has been used to acquire data on physical properties of the sea and on submarine wake investigations. The latest version of this vehicle, called the SPURV II, can reach depths of 5,000 ft (1,524 m). The SPURV II is 15 ft (4.6 m) long and weighs 1,300 lb (454 kg). Its navigation and control is performed by an operator through an acoustic link.

Researchers at the Naval Ocean Systems Center (NOSC) in San Diego have developed the AUSS (Advanced Unmanned Search System), which has a hull made from graphite cylinders with titanium endbells, allowing it to dive to very great depths of up to 20,000 ft (6,096 m). The AUSS uses an advanced acoustic communications system that transmits slow-scan T.V. pictures. This vehicle is specially designed and equipped for search of objects in deep-ocean areas.

Epaulard, built by the French company ECA, according to French government specifications furnished by NEXO (National Oceans Exploitation Agency), can perform photographic and bathymetric surveys on the sea bed down to 19,685 ft (6,000 m), and travels at a speed of up to 2.5 knots. Epaulard was demonstrated to the U.S. Navy in a mission to locate and photograph a World War II Douglas Dauntless bomber that was under

4,200 ft (1,280 m) of water. Although this vehicle is acoustically remote controlled, it also is able to follow some preprogrammed instructions. In looking for the Dauntless, the Epaulard followed its predetermined grid search without assistance from surface personnel until its obstacle-avoidance sonar detected an obstruction. The ROV relayed information on the obstruction to the operators on-board the ship and awaited further instruction. It was commanded to back off and make a new approach, which the Epaulard did successfully. This ROV was able to provide the navy with some very detailed black-and-white photographs of the four-decade-old aircraft. Most of this ROV's data is stored on-board for processing on the surface.

Radio and Laser Communications

Due to the inadequacies of acoustic communications, researchers are experimenting with other modes of signaling, including radio and blue-green lasers. Recent developments in Radio Remote Controlled Vehicles (RRCVs), whose signals travel at high speed and distances (as far as radio frequency is capable of), may prove to be an effective method of communicating with ROVs. The radio-controlled, high-speed DOLPHIN (Deep Ocean Logging Platform Instrumented for Navigation) operates near the surface at depths of 9.8 ft (3 m) to 16.4 ft (5 m) with speeds up to 15 knots in sea state 4 conditions.[1] This vehicle, built by International Submarine Engineering Ltd. of Vancouver, Canada, can transmit real-time computerized bathymetry data of the ocean floor to the mother ship via UHF (ultrahigh-frequency) radio waves. James McFarlane, president of International Submarine Engineering Ltd., has maintained that RRCVs would be made effective force multipliers by equipping them with electronic-warfare, antisubmarine-warfare, and mine-countermeasures equipment. RRCVs come in a variety of sizes and capabilities, depending on the mission objectives.

INDEPENDENTLY MINDED SUBS

Another approach to eliminating tethered ROVs is to not transmit command data at all and to make the vehicle fully autonomous. Such systems would have the artificial intelligence and sensors to enable the ROV to sense and interpret its environment, to locate and position itself, and to make real-time decisions as to the correct course of action.

U.S. Navy Programs

Several programs, sponsored by the military and private research organizations, are exploring autonomous submersible and submarine capabilities and technology. In 1975, the U.S. Naval Research Laboratory in Washington, DC, initiated the first autonomous ROV program, called the UFSS (Unmanned, Free-Swimming Submersible). This 5,420-lb (2,460-kg) vehicle, with a length of 20 ft (6.1 m), was designed for depths of 1,500 ft (457 m) and had an operating range of 125 nmi (230 km). The vehicle was initially used for ocean science information gathering and testing of the vehicle's laminar flow characteristics. Basic course and speed data were preprogrammed. There were plans to increase the UFSS's intelligence quotient by adding more precise navigational capabilities using inertial navigation with OMEGA or doppler sonar, and more onboard processing capability, pattern recognition, and artificial intelligence, as well as a mechanical manipulator. These enhancements were meant to allow the vehicle to assume additional functions such as the location of sunken submarines, mines, and other objects on the ocean floor. The one-eighth of an inch (0.3-cm) thick skin of the UFSS proved too susceptible to dents and full laminar flow was too difficult to achieve, resulting in the program's demise.

The U.S. Naval Ocean Systems Center (NOSC) in San Diego has developed a 400-lb (182-kg), 9-ft (2.75-m)-long, autonomous submersible named EAVE-West (for Experimental Autonomous Vehicle). The West differs from another ROV named EAVE-East, which will be discussed in a later paragraph. The EAVE-West has a gyrocompass, side-looking sonar, and a pipeline-follower subsystem that observes the magnetic field of the pipe, thus deriving steering signals. The ROV furnishes T.V. images to operators on-board a ship via a fiber optics link that is payed out from each end of the vehicle. The fiber link places no drag on the ROV, which can reach speeds of 5 knots. However, the EASE-West has only a one-hour operating time before the batteries need to be recharged.

The U.S. Naval Underwater Systems Center (NUSC) at Newport, Rhode Island, has built a 925-lb (420-kg) B-1 submersible, which is designed for optimum laminar flow, as reflected by its sharp nose, bulbous midsection, and small tail. It is outfitted with obstacle-avoidance sonar, a conformed array for reverberation study, side-looking sonar for self-noise and vibration study, and a relocation pinger. Its instrumentation records fifty-two channels of performance data and twelve channels of hot-film probe data.

The U.S. Naval Coastal Systems Center (NCSC) in Panama City, Florida, has built a 30-ft (9.1-m) scale model of an operational fleet submarine, which was designed to determine control characteristics under several transient conditions. An on-board computer is in complete control of the vehicle's performance. It also contains instrumentation to measure the dynamic motion of the vehicle, and a magnetic tape system for recording dynamic motion.

Commercial Autonomous ROVs

Two commercial, autonomous ROVs are furthering the technology in ways that may be applicable to military systems. The EAVE-East, built by the University of New Hampshire's Marine Systems Engineering Laboratory (MSEL) in Durham, has been adapted to a three-dimensional problem-solving system called SIMS (Structural Inspection Mission System). This allows the EAVE-East to locate, penetrate, and traverse an offshore structure, and then return safely. The AIMS (Arctic Inspection Mission System) will allow EAVE-East to map underneath arctic ice cover while taking bathymetric data, which are stored in its bubble memory for later display. MSEL is working on development of several other devices for EAVE-East, including a precise acoustic navigation system, capable of resolution of one-tenth of an inch (0.25 cm), for positioning the robot in a particular region and inside a structure; development of a miniaturized microcomputer for performing complicated tasks in less space; incorporation of a charge-coupled-device camera as a data-collection source; and a T.V.-image enhancement device that incorporates new algorithms, permitting redundancies to be eliminated.

The Autonomous Remote Controlled Submersible (ARCS), built by International Submarine Engineering Ltd., was developed specifically for surveying under the arctic's ice. This autonomous system can remain underwater for twenty-three hours at a depth of 1,200 ft (366 m), and travels at a speed of 5 knots. Using its five separate sonar systems and on-board computer, the ARCS can survey 2 mi^2 (5 km^2) on a preprogrammed course. It stores information on position, specific gravity, salinity, and depth. The internal software will also avoid obstacles detected by the sonar.

Mine-laying and Submarine Decoy Uses

In an appendix to the 1985 U.S. Defense Science Board report on "Military Applications of New-Generation Computing Technologies," board member Dr. Charles Herzfeld wrote that artificial intelligence technologies that have yet to be perfected could lead to a new breed of autonomous underwater vehicles for special naval applications. Herzfeld, who is a vice president with ITT Corporation, also suggested that autonomous systems endowed with vision and decision-making qualities could navigate for thousands of miles and penetrate enemy lines to perform missions too dangerous for manned ships. Two concepts were singled out in the report.

The first, as envisioned by Herzfeld, is an unmanned, autonomous, long-range underwater mine-delivery vehicle that could travel over thousands of miles to lay several dozen mines. In the process, the vehicle could also deposit sonar surveillance devices and trail towed-array sonars. Artificial intelligence would be needed for this vehicle to navigate over the vast distances and to make decisions as to where to lay the mines and how to cope with enemy ships and defenses, and environmental conditions.

Herzfeld's second concept involves using an autonomous vehicle to simulate the movement and signatures of submarines. Long-endurance decoys could mimic submarines leaving port. This capability would be especially useful for countering a Soviet attempt to establish a barrier in order to troll or intercept ballistic missile-deploying submarines. Such decoys could also be employed by attack submarines as a means to penetrate antisubmarine defenses. They could force enemy ships to converge in the wrong areas, or to use up their weapons and expendable sensors against apparitions.

There are certain advantages inherent in an unmanned submarine over that of manned ones. By taking men out of the submarine, the vehicle no longer needs the extra space devoted for comfort, warmth, and good air. The pressure inside the robot submarine can be equal to the outside sea pressure, and a vessel without frail humans at its control could take much greater punishment.

U.S. Research

Under a U.S. Navy-sponsored contract, research on a robot submarine designed for warfare in the twenty-first century is being conducted by the Marine Systems Engineering Laboratory of the University of New Hampshire in conjunction with Shenandoah Systems Company of Vienna, Virginia. This vehicle, known as the Long-Range Autonomous Submersible (LRAS), is conceived of as a high-payload vessel, ranging from 65 to 85 ft (20 to 26 m) long and 13 to 15 ft (4 to 4.6 m) in diameter. It could deliver a payload of 50 tons (45 metric tons) up to a distance of 10,000 mi (16,090 km). In the surveillance-mission profile, the LRAS would monitor traffic at strategic choke points with its TACTASS array. It also could analyze intercepted radar and communications data with its knowledge-based system processor. The LRAS would be ideal for the covert reconnaissance of heavily defended areas where extensive mining is expected or in locations inaccessible to manned submarines. The LRAS can also be used to probe enemy antisubmarine and surface forces to gauge their capabilities. This sophisticated ROV could also go on the offensive by laying mines and launching torpedoes or missiles at whatever target it has been preprogrammed to attack. Enemy-vessel acoustic signatures or other nonacoustic identifiers could be used to set the LRAS into its attack mode and, upon receiving further input, trigger its weapon systems.

The Westinghouse Defense Center's Oceanic Division in Baltimore is applying artificial-intelligence and expert-system techniques to an autonomous submerged vehicle concept for laying mines. Westinghouse has spent more than $1 million of its independent research and development funds on a knowledge-based control system for this program. This submersible would be able to go in and map a harbor, leave the harbor, figure out the optimum locations for mines, reenter the harbor to deposit the mines, and then depart the area. Gould and the Electric Boat Corporation are also developing a conceptual design for autonomous military submersibles and submarines.

The U.S. Department of Defense Advanced Research Projects Agency, in conjunction with the Robot Systems Division of the National Bureau of Standards (NBS), has initiated a project called SHARC (for Submarine with Hierarchical Autonomous Real-time Control). The program is aimed at demonstrating intelligent, cooperative behavior among several autonomous submersibles. DARPA will be using several of the EAVE-Easts of University of New Hampshire's Marine Systems Engineering Laboratory as the platforms to demonstrate the SHARC concept. The hierarchical real-time control system (RCS) architecture developed by the National Bureau of Standards will be incorporated within the EAVE-East. RCS consists of a modular hierarchically structured network of processors that communicate through a common memory. The hierarchical concept uses a top-down approach in which a global mission order is broken down into successive lower levels of planning and executing functions that send control signals to the vehicle's motors, actuators, and transducers. The vehicle will also be able to integrate sensory information into the processing, thus allowing for the stated variables to be tested against perceived reality. The ability to alter the RCS model hierarchy with sensory input is how mechanical learning takes place. The top levels of the hierarchy will also receive input from expert systems that will guide the individual group vehicles in strategy and tactics. One of the EAVE-East vehicles will be designated as leader and the others as follower vehicles. In 1987, DARPA and NBS will be testing how these vehicles communicate and share "thoughts" and sensory information.

British Research

British company Scicon Ltd. revealed their concept for a robotic submarine to meet next century's naval needs at the September 1985 Royal Navy Equipment Exhibition. The Scicon Patrolling Underwater Robot (SPUR) design (Figure 5.2) is of a 36-ft (11-m)-long vehicle that weighs 50 tons (45 metric tons) and is capable of diving to depths in excess of 19,629 ft (6,000 m). Its cruising speed will be 12 knots, but for its attack mode a 50 knot capability is projected. This high speed will be achieved by a closed-cycle internal-combustion engine. The SPUR's autonomous navigation will be derived from a digital hydrographic database. Scicon's viewfinder terrain

Figure 5.2. Scicon Patrolling Underwater Robot. (Courtesy of Scicon.)

modeling, in conjunction with a directional high-security echo sounder, will furnish the SPUR with comparisons of the actual depth against the database. An inertial navigation system will provide backup in areas where the bottom is flat and featureless.

Plans for the SPUR incorporate artificial intelligence architecture, allowing it to operate autonomously at sea for two-month patrols. Tactical weapons include its own torpedoes, as well as an integral warhead that explodes with the remaining fuel in suicide attacks. It will also be able to put wire around propellers for the covert impairment of enemy vessels. See Box 5.1 for a possible SPUR scenario.

A SPUR MISSION

Box 5.1 A squadron of SPURs will be transported to the area of operation by an auxiliary surface ship. Once reached, the SPURs will be launched over the ship's side from special underwater launch facilities, or by a helicopter when quick deployment is required. The SPURs will take up a patrol line, with spacing based on the sensor range. When the robot submarines run low on fuel or ammunition, or when they have been damaged, their internal monitors will activate a homing mechanism, returning them to their mother ship. When deployment is in a distant, hostile area near an enemy coast, operations are conducted more covertly with submerged deployment from disguised merchant vessels or from larger transport submarines.

M. W. Thring, the British robotics visionary, has suggested the idea of small robot submarines, carrying nuclear weapons, that would travel thousands of miles to locate enemy ports and coastal installations through dead reckoning and sonar navigation, and then proceed to destroy them with nuclear blasts. Other computer-controlled submarines would scour the oceans looking for hostile submarines, aircraft carriers, and large merchant ships. After detecting the vessels through sonar and magnetic sensors, the robots would use the method pioneered by Jules Verne's fictional *Nautilus*, that of ramming. The unmanned systems, however, would use small nuclear reactors to allow for ramming speeds of 93 mph (150 kph).

Summary

ROV missions have recently expanded from search, retrieval, and rescue tasks to performing mine location and destruction. The merit of building autonomous weapon systems is becoming more and more evident with the advancements in artificial intelligence architecture, robotics, underwater communications, and other related technologies. "Killer robots" will significantly affect how warfare is fought in tomorrow's oceans by conducting their own patrols and attacking with their own novel weapons. The survivability of manned systems will be in question.

REMOTE BOATING

The vulnerability of small surface vessels, as experienced in the Falkland Islands conflict, combined with spiraling manpower costs has caused naval planners to reconsider an approach used in World War I, that of deploying remotely controlled surface boats. Unmanned boats have been used for some time as targets for surface guns and underwater weaponry. New missions for these vessels include use as a coastal patrol craft, as decoys for larger ships, to lay and dispose of mines, and to conduct scientific studies of the ocean's salinity, dynamics, and geophysical aspects.

Aldebaran Remote Boat and Tracking System

Pacific Aerosystem Inc. of San Diego and Meteor SPA of Italy have jointly developed the Aldebaran Remote Boat and Tracking System. The system uses the Excalibur boat originally built by Meteor, which is a modified 32-ft (9.7-m) ocean racer. It is constructed of fiberglass and wood, and is capable of reaching 48 knots in sea state 4 and up to 60 knots in smooth-sea conditions. The Excalibur may be operated in three different modes: a remote-controlled mode, a preprogrammed mode, and a remote-navigation mode.

The remote-controlled mode involves operator use of the Alamak system, which allows him to remotely start and stop the boat's engines, specify the boat engine's revolutions per minute, and control the optional T.V. camera's pan, zoom, and focus. The operator can let the vessel be controlled by an autopilot, or can manually specify the rudder order angles to command the boat's heading. The Alamak control console enables the operator to track and plot the Excalibur's range, to display telemetry data, to activate mission programs, and to self-test the boat's components.

The second mode of operation involves proceeding along a preprogrammed course. The Excalibur boat can follow a number of repetitive patterns, including figure-8 and racetrack maneuvers. Up to twenty-eight waypoints are available to the mission planner for designing a very complex pattern to conform to a harbor's features or other configurations. While in this mode, the program can be interrupted and control returned to the operator, unless the boat's computer is programmed to ignore all radio command signals as a way of maintaining security. In the third mode of operation, the Excalibur can use on-board navigation aids such as satellite navigators to maneuver it to the desired location.

The Aldebaran system has a multitude of applications, ranging from training derived from its use as a target to reconnaissance and strike missions. Its ability to patrol seaways and harbors without the consideration of personnel fatigue makes it attractive to the military. The Excalibur's low radar cross section allows for minimal detectability; it is well suited, therefore, for covert operations. Meteor had previously sold fifty of its Excalibur boats to Libya. This has caused some headaches for the U.S. Sixth Fleet, because the Libyans have packed their remotely controlled boats with explosives and may have given some to terrorist organizations. The U.S. Navy has also experimented with the Excalibur boat and the Alamak control system in performing a barrier patrol task during an exercise held off the coast of California.

The Seaflash

The Seaflash series 300 remotely controlled surface craft (Figure 5.3), which is built by British firm Flight Refueling Ltd., is in service with the Royal Navy and several other fleets around the world. It has a 28-ft (8.6-m)-long glass-reinforced plastic hull, with a beam length of 7 ft, 3 in. (2.2 m) and a height from its waterline to its command aerial of 19 ft (5.8 m). The petrol-driven V-8 engine allows the robotic boat to reach speeds of 30 knots (20 knots at sea state 3). It is controlled by a radio data link with six or more functions, and has a range of 10 nmi. The Seaflash also has an optional on-board manual control, if warranted.

The primary function of the Seaflash is to provide target practice for

Figure 5.3. Seaflash remotely controlled surface craft. (Courtesy of Flight Refueling.)

surface gunnery and underwater weapons. Other missions include serving as an electronic countermeasures decoy via the use of infrared radiation or by emitting chaff or smoke; as a simulator for fast patrol attacks; as a craft for reconnaissance and the remote detection of radiation or toxic substances; and as a platform for mine detection and detonation. Maintenance and operator training for the Seaflash are quite minimal, and it can be easily hoisted aboard larger vessels and then stowed.

Other Concepts

Dr. Jasper C. Lupo, program manager in the Tactical Technology Office of the U.S. Defense Advanced Research Projects Agency, has advocated a concept called "Sharkpack," which features unmanned frigates and destroyers to escort merchant vessels traveling in convoys. These robot surface craft would be remotely controlled by a military commander located with the convoy, and would guard against attacks from enemy submarines and patrol vessels.

Another concept envisions using unmanned hovercraft-type vehicles to engage enemy surface fleets in quick hit-and-run type attacks, or in amphibious assaults on enemy-held beachheads (Figure 5.4). Several such robotic vehicles could be controlled by an accompanying manned hovercraft, which would stay safely in the rear.

Any adoption of these revolutionary concepts, involving the heavy use of robotics, might alter naval warfare as much as any previous development, even the adoption of steam-driven ships and submarines. There are very successful models for this required revolution in thinking. One must only carefully study the fourth environment within which military action might take place—that environment above the earth's atmosphere commonly known as space. The most important uses for military robots have already taken place in space as earthlings rely upon hundreds of unmanned reconnaissance, communications, and navigation satellites currently in orbit. Several countries have demonstrated the feasibility of unmanned space probes that have successfully explored millions of miles of space. Since exploring space may indeed be man's "last frontier," the recent entry into space means there are fewer vested and bureaucratic interests involved. There thus may be less resistance to using unmanned vehicles to perform missions in space. Finally, the space environment is hostile to humans and often demands rapid calculation and decision-making beyond the capacity of unaided human beings. For all these reasons, we will likely see above the earth's atmosphere the most innovation in the fielding of robotic weapon systems. Space provides a unique perspective and ability to monitor and intervene on the earth below and will become increasingly important to the security of the planet.

Figure 5.4. Artist's concept of several unmanned hovercraft being remotely controlled by manned hovercraft. (Courtesy of U.S. Army.)

NOTES

1. Schwartz, M., Ed. *The Encyclopedia of Beaches and Coastal Environments.* Hutchinson Ross Publishing Company, Stroudsburg, PA, 1982, pp. 722 and 723. The U.S. Navy Hydrographic Office uses the Douglas scale in which to measure the sea state condition, which is the condition of the ocean surface in roughness and wave height. Sea state 4 indicates that conditions are rough, with the height of waves being 5 to 8 feet.

6 Space-based Robotics

THE HIGH GROUND

A key advantage in winning a battle is to hold the "high ground." From this commanding position the enemy can be more easily observed, and if need be, targeted. In contemporary warfare, securing the high ground means controlling space. Communications, navigation, photo intelligence (phoint), electronics intelligence (elint), signal intelligence (sigint) and strategic weaponry are in large part dependent on operations in space. The competitive edge for controlling this high ground over the next few decades and beyond will go to the nations with the most effective military space stations and space-based weaponry. Robotics technology is an essential ingredient for producing these extremely complex systems. People will continue to function in the heavens during peacetime, but the warfare environment of nuclear-pumped X-ray lasers, electromagnetic rail guns, and charged-particle beams will be very deadly for humans. The leaks resulting from the release of streams of rays, beams, and particles would furnish a deadly dose to any person on-board the space-weapon systems. Unmanned, robotic space vehicles are the only viable option for space weaponry.

Of all the hardware circling the Earth, at least three-fourths has a military function; the vast majority of it is unmanned. Military satellites are vital to both strategic and tactical warfare. Space-based robots are the "eyes and ears" of defense establishments, which are becoming increasingly more dependent on them.

SPACE STATIONS AND THEIR ROBOTS

Even manned-space-station programs will entail a vast amount of robotics and automation. Although space station operations will include civil applications, their functioning will be vital to national security concerns. The

Soviet Union already has a rudimentary space station. *Soyuz* spacecraft are used to transport cosmonauts and supplies between the Earth and the *Salyut* space station. The U.S. space station program is much larger in scope and involves the use of the space shuttle to transport materials and astronauts for its development. Potential military purposes for future space stations include the deployment of weapons in space as well as the stationing of special space-based troops (human or robotic), which can be used against enemy satellites, space stations, strategic defense space platforms, and even ground targets.

A 1984 study by the U.S. National Aeronautics and Space Administration (NASA) evaluated alternative space station designs and concepts. It concluded that the space station and the manned program in general should incorporate an extensive use of advanced automation, robotics, and artificial intelligence. Some of this technology is currently available for space station use, but obtaining the bulk of it will require the commitment of substantial research and development dollars devoted to maturing robotic technology.

The NASA study highlighted several major benefits derived from the use of robotics and automation. Astronauts would be viewed more as managers on behalf of station users than as operators carrying out routine functions. The use of expert systems and advanced machine intelligence would enable station personnel to detect, diagnose, repair, and recover from abnormal situations without dependence on ground-based mission control. Astronauts would thus obtain greater autonomy and be able to make decisions at the station.

Robotic vehicles and systems would lessen humans exposure to hazardous situations. For example, extravehicular activities (meaning outside of the space station), including fueling tasks and servicing satellites in very high (geosynchronous) orbits, may subject the individual to harmful radiation.

Using robotics would make the space station more adaptable and flexible to commercial ventures and their manufacturing applications. Production would run at peak efficiency from the use of unmanned systems, which would also result in a lower cost of operations. Spin-off economic benefits may be derived from the lessons learned from space robotics and automation advances, and robotic technology developed for space may be applied to terrestrial hazardous settings, such as under the ocean, in nuclear power plants, or on the battlefield.

Company Studies of Space Applications

In an effort to gauge how robotics and advanced automation systems can best be applied to the space station, NASA contracted out to several aerospace companies to study specific applications areas. Boeing examined the human–machine interface, and in so doing determined that it was technically feasible to develop a rudimentary extravehicular (EV) robot by the mid-1990s, and a more sophisticated EV robot by 2010. The EV robot, slightly larger than a human, would be used to relieve astronauts of routine and hazardous tasks as well as increase extravehicular activity around the space station.

Conceptual designs for gallium arsenide crystal and microelectronic-chip manufacturing facilities that incorporate automated machinery were produced by General Electric. G.E. found that robots and teleoperators built for space-based facilities should be constructed of much lighter materials than exist in Earth-bound machinery in order to compensate for microgravity environment kinematics and dynamics. A major challenge is to develop "intelligent" tools to maintain, repair, and refurbish production equipment.

Hughes examined automation as it affects subsystem control and mission operations, with an emphasis on developing concepts for autonomous operations. In addressing electric power, thermal control, and communications, Hughes concluded that automatic speech recognition and synthesis should be incorporated in the human–machine interaction for space station command and control. Automated machinery should be able to detect space station faults, isolate them, and find alternative actions in order to recover from failures.

The required automation for the assembly, construction, repair, and modification of a year-2000 space station was researched by Martin-Marietta. Because their study found that automation technology will grow at a rapid pace, they recommended that the space station design be inherently adaptable to accommodate this evolution. For example, access corridors, berthing ports, and work-site "rest points" should be designed with enough room and support equipment for the addition of manipulating devices as they are developed. Space-based construction was foreseen by Martin Marietta to require a great deal of human involvement at first, which, however, should decrease over time, until humans serve only in a managerial or contingency capacity.

TRW tackled automated satellite servicing from the space station, which involved support to a low-Earth-orbit satellite and an orbiting materials-processing facility, handling payloads, and servicing a geostationary satellite by a recoverable Orbital Transfer Vehicle (OTV). TRW concluded that accelerated development in automated hardware that uses telepresence (remote-control) equipment is needed for satellite servicing, but human involvement will still be required to handle the more diversified or unforeseen events.

NASA's Robotic Vehicles

Space Tug

NASA has near-term plans involving the introduction of several robotic vehicles and systems for space station operations. The Orbital Maneuvering Vehicle (OMV), commonly referred to as the Space Tug, is a remotely piloted craft designed to increase the operational range provided by the space shuttle. Carried aloft by the shuttle, the OMV could then be launched up to an additional 1,400 mi (2,254 km) to retrieve satellites in very high orbits. These satellites could either be serviced by the shuttle crew or returned to Earth for repair. The OMV then places the satellites back into their proper position. The U.S. Department of Defense also acknowledges the great savings that would be derived from the resulting extended lifetime of their satellites.

This robotic vehicle is designed to be maneuvered by a hand controller manipulated by an operator who is at a ground station. LTV, Martin-Marietta, and TRW all competed for the contract, which NASA awarded to TRW in July 1986. NASA hopes to have it in operation during the 1990s. The Space Tug will also operate in conjunction with the space station to remove space debris and to perform various construction tasks.

The OTV

Another developmental unmanned vehicle is the Orbital Transfer Vehicle (OTV). The OTV is conceived as an upper stage that can ferry payloads into geosynchronous orbits as high as 22,300 mi (35,903 km) above the Earth. It would then return on its own for later reuse. OTVs that would remain in space and those that would travel both to and from the Earth are being considered by NASA. The U.S. Department of Defense and NASA have both shown strong interest in a system named DARTS (Delta Astrotech Reusable Transfer Stage) that is being developed by Astrotech International Corporation of Pittsburgh. Astrotech estimates that full development of the reusable type of OTV would require about thirty months from the point of funding. Astrotech would handle the program management and would use McDonnell Douglas Corporation as the prime contractor for system development. The U.S. Air Force awarded a contract to Boeing Aerospace to build Inertial Upper Stage (IUS) boosters, another form of OTV, to propel satellites from the orbit attained in their delivery by the space shuttle into geosynchronous earth orbits. The IUS, however, is not designed to be reusable.

The TSU

A vehicle that resembles NASA's Manned Maneuvering Unit, although without the astronaut, is the Telepresence Services Unit (TSU), which is configured to perform maintenance at the same level as a man could in space. The TSU was designed by the Massachusetts Institute of Technology in conjunction with NASA. It is a square-shaped vehicle with two large manipulator arms on the front and two smaller anchor arms on its sides. Below these arms is a rack of tools and spare parts. The TSU will be dependent on the OMV for transportation to and from the work sites, for battery recharging, and for storing large parts.

Robotic Arms

NASA plans to equip the space station itself with several robotic arms resembling that carried by the space shuttles. These manipulators are being designed to attach to a mobile platform, allowing them to reach all over the space station to assist both in its construction and in its operations. The robotic arms may be mounted on a platform that rolls across the station's main structures, or one that uses end effectors to walk across the station. As with the space shuttle, Canada is assuming the lead role in developing robotic arms for the station. This effort, known as the Integrated Service and Test Facility (ISTF), involves equipping the manipulators with tactile and vision sensors to provide for more autonomous, as well as dexterous, handling of payload refurbishment. Tedious detail work, such as unscrewing bolts, will be able to be performed by the robotic appendages, thereby allowing the astronauts to function as managers and supervisors. Spar Aerospace Ltd.'s Remote Manipulator Systems Division at Weston, Ontario, is the prime contractor. They envision the ISTF as servicing the OMVs and OTVs as well as the space station, by the turn of the century.

Military Services' Research

In February 1986, the U.S. Air Force Systems Command announced the initial results of their Project Forecast II program, which is aimed at identifying and using new and emerging technologies and systems to enhance air force warfighting capabilities into the twenty-first century. Unlike many military long-range-planning efforts, the air force is devoting significant resources, approximately 10% of the service's science and technology budget, over a six-year period to finance the project. An estimated $2 billion will be spent on Project Forecast II technologies and systems. Among the technologies that the project plans to thoroughly examine is robotics. It calls for exploring how remote-controlled robots could assume tasks performed by astronauts, in order to reduce the risks encountered in space flight. The air force is especially interested in teleoperators, through which the human operator could sense in three dimensions whatever the robot is experiencing (Figure 6.1). They want a system with good eyes and strong arms, but virtually no brainpower. Space applications for such teleoperators include placing them in geosynchronous space stations to repair satellites, and eventually in lunar bases.

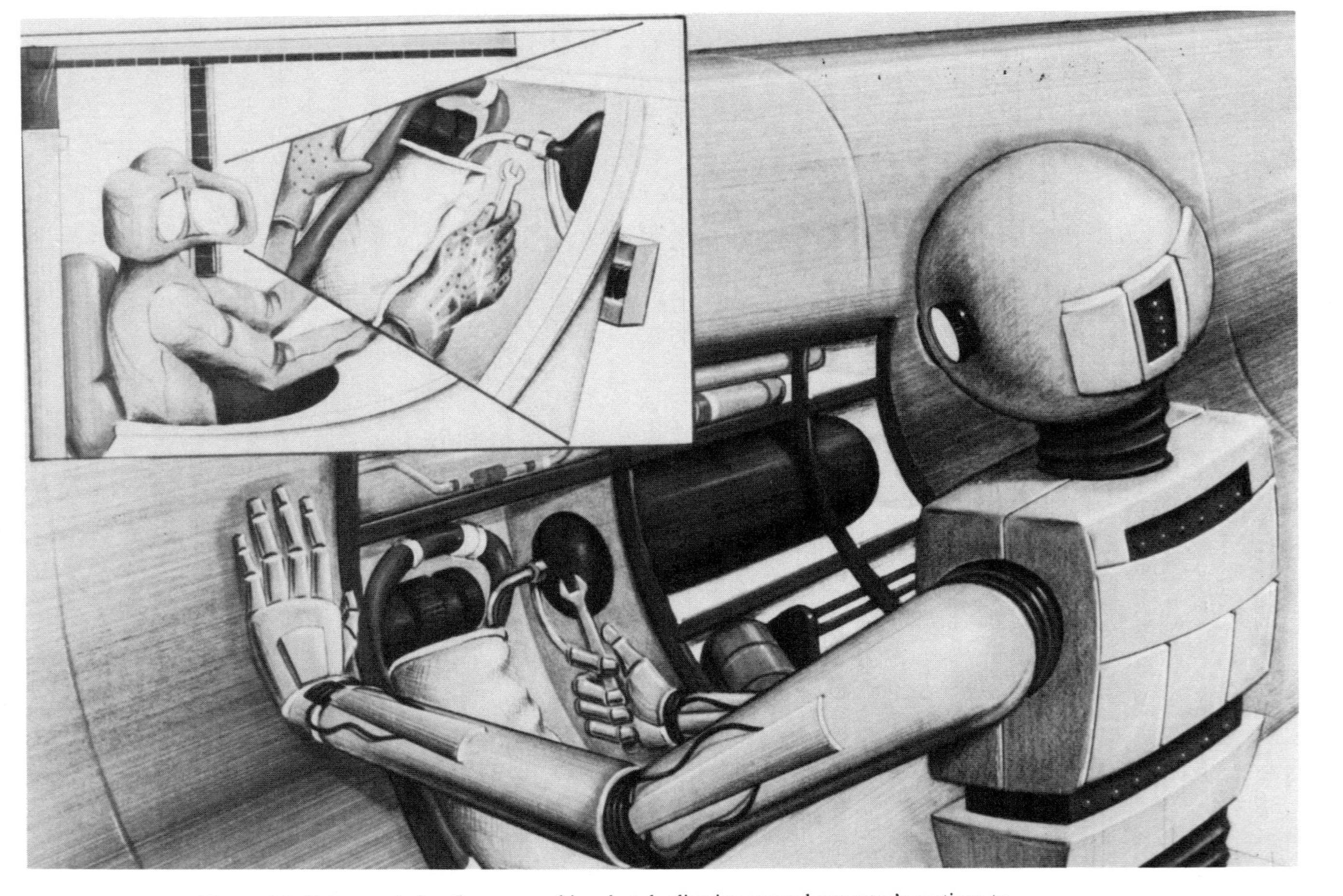

Figure 6.1. Teleoperated anthropomorphic robot duplicating ground-operator's motions to repair space system. (Courtesy of U.S. Air Force.)

The ability of scaling robotic operations up or down offers some interesting opportunities. This would allow the performance of heavy construction at 10 or 100 times the rate of a human operator to which the robots are slaved. It would also allow a human to manipulate a large circuit board while the robot performed the same manipulations on the microchip level. Besides space uses, the air force believes that anthropomorphic teleoperators can also be useful in performing aircraft repair and replenishment in NBC environments.

The U.S. Naval Oceans Systems Center at San Diego, California, has already been working on an anthropomorphic system with applications for space and other hazardous settings. In this system, the operator receives input from various sensors that give him the feeling of "remote presense." The teleoperator consists of a hydraulically driven head with three degrees of freedom that is mounted on a torso that also has three degrees of freedom. Attached to the torso are two manipulators resembling arms and hands that have seven degrees of freedom. A pair of T.V. cameras and a binaural hearing system furnish the operator with remote vision and hearing. The operator's motions control the teleoperator's motions. The operator's head, torso, and arm positions are monitored by potentiometers, which then signal the teleoperator, whose mechanical exoskeleton and manipulators adjust accordingly.

STRATEGIC DEFENSE ROBOTS

The speculative fiction of the early twentieth century featured rocket ships flying across the heavens, pitting Earthmen against aliens. Flying duels fought with "death rays" of great destruction typically lead to the "zapping" out of existence of the evil enemy spacecraft. The infant film industry introduced the masses to such space heroes as Buck Rogers and Flash Gordon. Children whose imaginations were shaped by these visions have become scientists, military leaders, and even U.S. presidents, people who are now capable of converting these dreams into reality. As the end of the twentieth century approaches, mankind has entered the "Star Wars" era. Unlike the popular George Lucas movie, however, the spacecraft emitting lasers and charged particles will be devoid of human occupants. Inanimate, unemotional objects equipped with the greatest computer-processing capabilities ever devised by mankind will be making the key target selection and firing decisions.

Strategic Defense Initiative

President Ronald Reagan's March 1983 speech placed into policy a program dedicated to making nuclear weapons obsolete by intercepting and destroying strategic ballistic missiles before they reached their targets.[1] All

U.S. ballistic missile defense (BMD) research programs were placed under a separate Strategic Defense Initiative (SDI) Command.

The U.S. SDI program involves a three-layered defensive system (see Figure 6.2). The majority of intercontinental ballistic missiles (ICBMs) would be targeted for destruction soon after being launched. During this boost phase, which typically lasts three to five minutes and encompasses the time period from launch to burnout of the final stage, the rocket boosters provide a very visible and vulnerable target. Those remaining would be reattacked during the midtrajectory phase, at which point the ballistic missile has entered space prior to its descent. This phase may last up to twenty minutes. The surviving warheads would be attacked while reentering the atmosphere by ground-based antiballistic weaponry. This terminal phase lasts less than a minute. It is during the boost and midtrajectory phases when robotic space-based weapon platforms will be the primary defensive systems.

The actual vehicles are still early in the conceptual stage, and various configurations have been proposed. Some Star Wars advocates have suggested a system of fifty space battle stations to protect the entire United States from Soviet missiles. Each space battle station would contain a highly sophisticated computer system capable of detecting, identifying, tracking, and aiming a powerful laser at a number of missiles, and then sequentially hitting them in rapid succession. A ground-based command post would contain a main computer to control each battle station. Early warning satellites capable of detecting missiles soon after launch would relay this information to the command center, which would in turn activate the space battle stations.

Lasers and Charged-particle Beams

Several varieties of lasers are being studied for use on the battle stations (Figure 6.3). These include chemical and X-ray lasers. Chemical lasers obtain their power from the reaction generated by the mixing of two gases, such as hydrogen/fluorine, oxygen/iodine, and deuterium/fluorine. One SDI initiative involved developing a 2-MW hydrogen-fluorine laser to demonstate the feasibility of building a 25-MW space-based laser. The X-ray laser is perhaps the most controversial SDI approach. It involves placing up to fifty lasing rods around a low-yield nuclear explosive. These rods would be pointed at the incoming missiles, and upon detonation of the nuclear warhead, they would emit intense X-ray pulses. X-rays would damage the missile's electronics, as well as generate shock waves that could destroy the missile's skin.

Excimer and free-electron lasers would be too large and consume too much power to be based in space. Another defensive system concept, however, envisions the use of space-based vehicles equipped with a giant

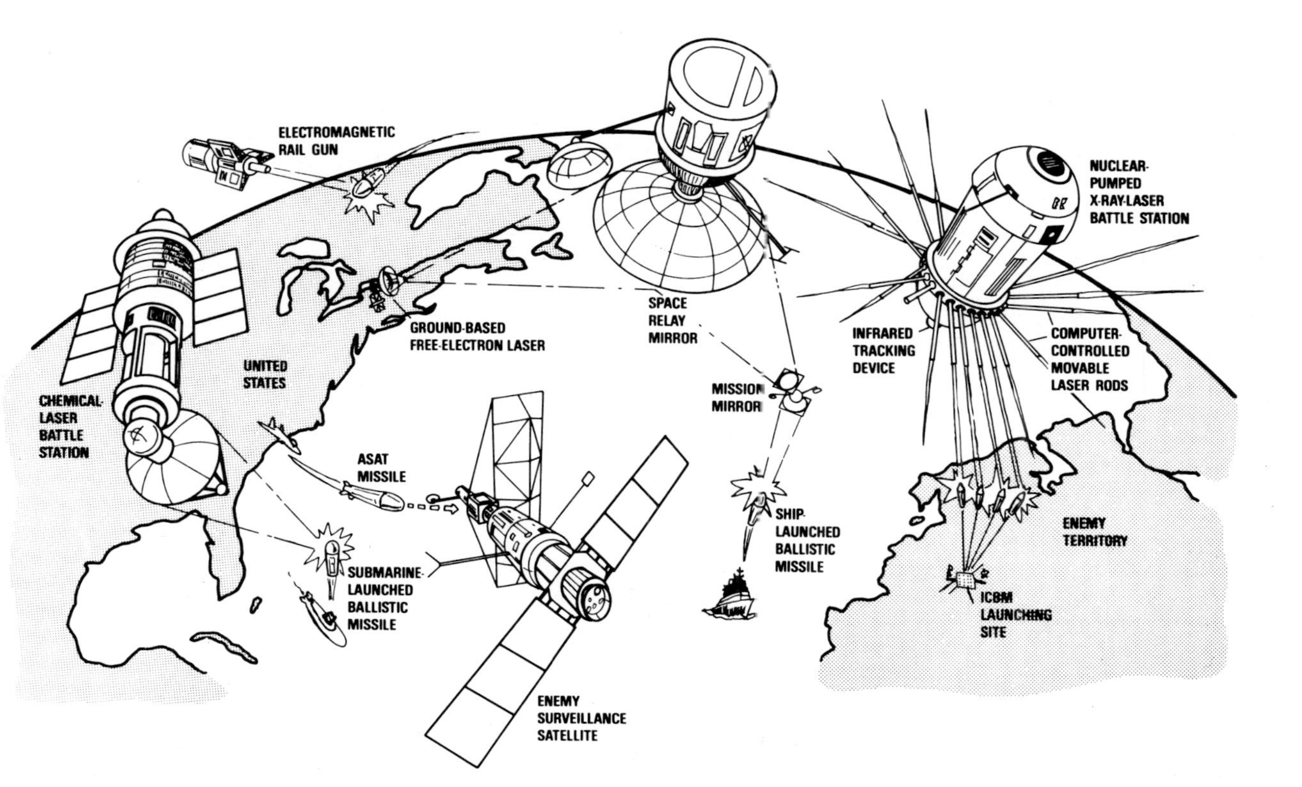

Figure 6.2. Sketch depicting various conceivable space-based robotic antimissile systems. (Courtesy of Titan Corporation.)

Figure 6.3. Artist's depiction of a robotic space-based laser system. (Courtesy of Lockheed.)

mirror 13 ft (4 m) in diameter to redirect ground-based free-electron and excimer laser beams to targets in space.

The use of charged-particle beams is also being examined in the SDI program. Comparable to a controlled bolt of lightning, charged-particle beams involve accelerating subatomic particles to velocities approaching the speed of light via electric fields in particle accelerators. Accelerators producing such beams must be light enough to be economically placed in space. One difficulty facing charged-particle beams is that the Earth's magnetic field will bend them over long distances. A possible solution to using these beams in low-Earth orbit is to have a laser punch a hole through the rarefied near-space environment, through which an electron beam could be fired. Theoretically, the positively charged gas ions would provide an electrostatic restoring force that would compensate for the effect produced by the Earth's magnetic field.

Kinetic Energy Devices

A traditional approach to destroying a target is to hit it with an object traveling at a speed greater than that of the target. Kinetic-energy space-based vehicles are intended to destroy missiles by collision, using the kinetic energy inherent in moving objects. One possible device is the electromagnetic rail gun, which operates by passing a powerful electrical current, up to several megaamps, through two parallel rails. A projectile fits between the rails and makes electrical contact across them. Upon starting current flow, the projectile is propelled along the rails. Experimental rail guns have accelerated 3.5-oz (25-g) particles up to speeds of 5 mi/sec (8.6 km/sec) and have penetrated steel plates over one-fifth of an inch (0.5 cm) thick. Larger projectiles weighing 10.5 oz (300 g) have reached velocities of 2.6 mi/sec (4.2 km/sec). Scientists have calculated that a projectile velocity could be increased to 62 mi/sec (100 km/sec) with rates of fire up to sixty shots per second. An orbital gun system using charged-particle beams is being developed by the SDI office to defend against Soviet antisatellite (ASAT) weapon systems.

Another kinetic-energy approach uses heavier projectiles equipped with their own guidance systems. The "high frontier" concept proposes a network of 400 space-based robotic vehicles, each outfitted with fifty interceptors. The interceptors would consist of miniature homing devices with their own propulsion systems. One air force official estimated that an effective antimissile system might require up to 3,200 rocket-carrying battle stations located in forty different orbits. Satellites carrying radar and other sensors would be needed to augment the battle stations.[2]

Size and Structure of SDI

There is no real argument in the United States concerning the efficacy of spending research and development dollars on space-based robotic vehicles. If nothing else, this would guard against technological surprise. The real debate is on how large the program should be, and if the program should be structured to try to protect the whole U.S. population and make nuclear weaponry obsolete as envisioned by President Reagan, or whether it should be used to protect only vulnerable elements of the U.S. strategic forces, such as ICBM silos and command and control nodes. A point defense would be much cheaper and technologically more feasible to obtain than one necessary for defending the nation as a whole. Point defense of vulnerable strategic forces could result in a more credible retaliatory threat and serve as a deterrent to war. This would, however, maintain the Mutually Assured destruction (MAD) doctrine of holding each country's populace as hostages, something many SDI proponents are against. On June 18, 1986, the U.S. Senate Armed Services Committee passed a policy statement that the SDI program should continue research on the potential for population defense. The major emphasis, however, should be on developing survivable and cost-effective defense options to protect the retaliatory forces and their command and communications systems. U.S. Defense Secretary Casper Weinberger countered the Senate committee's vote by stating that the administration's policy would continue to be geared toward protecting people, not just the missiles. He pointed out that a key factor in SDI's success would be to attack Soviet missiles as they take off, which negates any ability to distinguish between countervalue- or counterforce-aimed missiles. Weinberger did concede, however, that they were having success with prototype weapons that would be useful for the terminal-phase close-in defense of missile silos from warheads reentering the atmosphere, which is the approach advocated by the Senate committee. Political considerations, technological progress, the state of the economy, and arms-control negotiations will ultimately shape the evolution of the SDI program.

TRANSATMOSPHERIC VEHICLE

In his 1986 State of the Union message, President Reagan reaffirmed U.S. commitment to the space program in light of the tragic space shuttle *Challenger* explosion. He used the occasion to announce a shuttle follow-up program known as the transatmospheric vehicle (TAV) or "space plane."[3] This vehicle is capable of flying from the Earth through the atmosphere into space and back, and can travel at hypersonic velocities (Figure 6.4). Such a vehicle has both military and civilian applications. The primary military mission for the TAV, known in the military as X-30, is to launch SDI-related

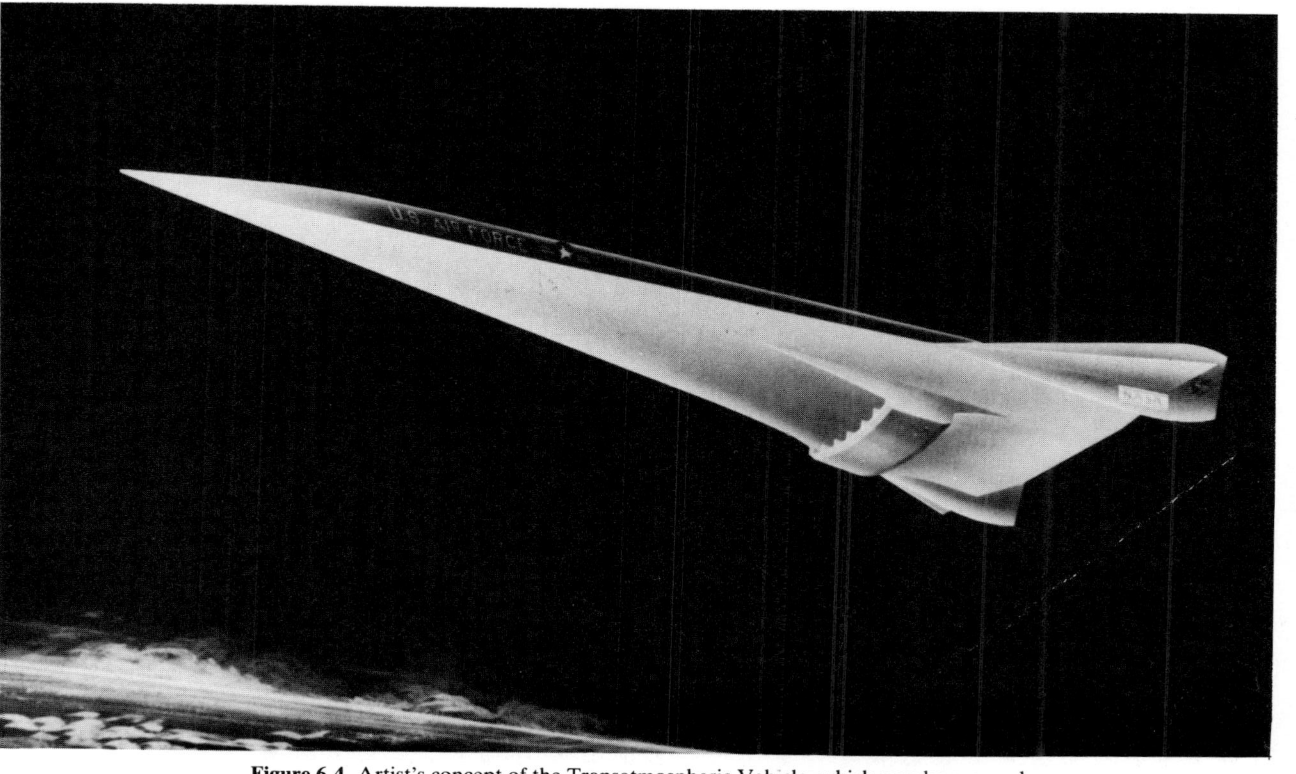

Figure 6.4. Artist's concept of the Transatmospheric Vehicle, which may be manned or unmanned depending on the mission. (Courtesy of NASA.)

payloads into space. It is estimated that the TAV will be able to carry a much greater payload than can the shuttle, and at lower cost. According to Robert C. Duncan, director of the U.S. Defense Advanced Research Projects Agency (DARPA), the TAV is also being considered as a long-range air-defense interceptor, used to attack incoming ballistic missiles.

Although the TAV is primarily perceived as a manned system, a great deal of automation and robotics technology will be incorporated into its design. Unmanned vehicle proponents are advocating that the TAV be built with the ability to be flown autonomously, by remote control, or manually. With this feature, consideration of the complexity and hazardousness of each particular mission could determine whether the vehicle should be manned or unmanned.

The British National Space Agency has begun development of a comparable TAV named the Horizontal Take-Off and Landing (HOTOL) craft. British Aerospace and Rolls Royce are performing the bulk of the conceptual and feasibility studies. Current emphasis is for unmanned missions by the HOTOL to deploy military and commercial satellites.

PLANETARY ROVERS

NASA has some ambitious plans involving development of unmanned rovers by the year 2000. These rovers, which would transport scientific instruments some 1,200 mi (1,930 km) across the lunar surface, will be equipped with an automated lab to examine soil, and with T.V. cameras to help map in detail the areas covered. Such information will be necessary in setting up the first moon base, which the United States has tentatively planned for the year 2007, the fiftieth anniversary of the beginning of the Space Age. In a similar program, NASA's Marshall Space Flight Center has developed a Martian rover concept known as the Elastic Loop Mobility System (ELMS). The ELMS is an elastic-loop-tracked device that would replace the three landing pads used on spacecraft such as the Viking landers. The ELMS would be equipped with stereo cameras, manipulator arms, and various scientific instruments to study the Martian surface.

NOTES

1. Barnaby, Frank. *Space Weapons*. Gallery Books, New York, 1984. p. 7. Barnaby excerpts President Reagan's "Star Wars" speech of March 23, 1983. He quotes the president as stating: "I call upon the scientific community who gave us nuclear weapons to turn their great talents to the cause of mankind and world peace; to give us the means of rendering these nuclear weapons impotent and obsolete."
2. Hiatt, Fred. "Space Launch Needs of SDI Are Estimated: 600 Liftoffs Seen in a 3-Year Period." *The Washington Post,* June 27, 1986, p. A24. Hiatt excerpts the *Military Space* newsletter, which quotes air force Col. William Zersen on battle

station requirements. Zersen later told Hiatt that these were only preliminary "ballpark" estimates.

3. Hiatt, Fred. "Space Plane Soars on Reagan's Support." *The Washington Post,* February 6, 1986, p. A4. Hiatt quotes President Reagan's February 4, 1986, State of the Union Address: "And we're going forward with research on a new Orient Express that could, by the end of the next decade, take off from Dulles Airport, accelerate up to twenty-five times the speed of sound, attaining low Earth orbit or flying to Tokyo within two hours."

7 Robots on the Future Battlefield

ROBOT RETROSPECTION AND PROJECTION

Around the world, in almost a dozen nations, a vast amount of research and development is being done on both civilian and military robotics. In this book, we have catalogued several hundred major research and development programs, some funded by governments, others undertaken by entrepreneurial companies. Robotic systems are already playing a major role in many areas of our life, particularly in assembly line manufacturing. The mechanical abilities of robots will be improved, their mobility will be enhanced, and they will be linked with more sophisticated sensors, data processing capabilities, and communication networks. We asked a question at the outset of this book—Will there be robots on the future battlefield? This book has clearly answered this question. In the twenty-first century, there will be battles, if not wars, fought without men and between robots.

Compiling the enormous amount of data on developments in robotics and their application to military issues was a monumental task. However, it is dwarfed by our goal in this chapter—to challenge, define, suggest what all of this means. We will not presumptuously claim to predict accurately what precisely will happen in the twenty-first century. Rather, in order to bring some clarity and focus to this discussion of hazy and unfocused future, we will examine why the lethal battlefield will encourage the increasing use of robots in warfare; why teleoperated systems will predominate in the near term and will be followed by autonomous robots and multi-environment robots; which systems are likely to emerge first in the environments of the ground, air, sea, and space; how robots might be used by or against terrorists; and why the inevitable development of military robots will stimulate major political and social debate, and resistance to these trends. Finally, our forecast of military robotics will also be accompanied by a caution that humans must take measures to ensure that mankind remains in control of these metal creatures capable of great devastation.

BATTLEFIELDS UNFIT FOR HUMANS

If we must be concerned about controlling robots, how can we be so sure that conditions will lead to their introduction on future battlefields in significant numbers? By examining the potential economic costs, territorial devastation, and lethality of contemporary warfare, one rapidly appreciates why the proliferation of robotic weapon systems will be a logical outcome. The use of unmanned vehicles is generally accepted as being advantageous over manned systems when the following criteria are met:

1. When the lethality of the mission is too great or when our cultural norms prohibit us from committing soldiers to suicidal missions. Robotic vehicles may survive extremely toxic or explosive environments and, if destroyed, only an expensive piece of equipment needs to be replaced.
2. When human resources need to be diverted to other priorities. Robotic systems can free essential manpower to perform higher priority missions by taking on less complex, redundant missions.
3. When the overall efficiency and effectiveness of a task can be better accomplished through automation.

Battlefields are becoming too intolerable for humans to carry out their missions. If we do not already have, we are very close to having the technological means to deny any piece of "real estate" to the enemy by employing NBC agents. Political considerations and the fear of retaliation in kind have prevented this from being tested in the past. With the advent of precision munitions, lasers and kinetic energy and charged particle weaponry, our ability to hinder human occupation and movement in battlefield areas will be even greater. What restraints will be maintained on tomorrow's battlefields are difficult to predict, but it may very well be that only unmanned systems that are immune to toxic substances and have greater resistance to impact and heat weaponry will have a future there. Unmanned rovers have already proven quite adept at performing tasks in hazardous environments such as nuclear power plants, outer space, under the sea, and in factory settings unfit for human labor. Such devices can also survive in hostile battlefield conditions.

Efforts are being made to improve the human's chance for battlefield survival. Researchers have made great progress in producing lighter and more effective clothing materials and designs to protect personnel from NBC contaminants. Even so, performing essential military tasks while wearing the latest in protective apparel is still a very unwieldy act that significantly degrades the timeliness and quality of the work. The donning of protective masks and hood, overgarment, gloves, and overboots incurs too many operational costs. Although bulkiness has been reduced, the garments still trap heat and moisture, limiting their use to relatively short periods of time. In desert operations, only a few minutes of protective clothing wear is possible. The protective ensemble reduces touch, smell, sight, and sound and makes many basic human needs difficult to accomplish. Eating, drinking, and waste elimination are tedious. Electronic component repair or even typing may be quite difficult, if not impossible, under such circumstances. The damage to protective gear from jobs typically associated with "bruised knuckles," such as automotive repair, becomes quite deadly to an individual who tears his/her glove in a contaminated environment. The use of robotics and automated machinery may be the only way to perform missions safely, efficiently, and effectively in the NBC environment. Repairing equipment, performing vehicle maintenance, fighting fires, and ministering to the wounded are quite difficult under normal circumstances; they are almost impossible in the presence of deadly toxic agents or incapacitants. These functions will be eventually accomplished in the NBC environment by removing humans to safety, and performing the missions with teleoperated and autonomous vehicles.

The employment of NBC weaponry to achieve speedy victory is feasible, although it is far more likely to be used to weaken the opposition. This is accomplished by forcing the opponent to devote too much time and resources to the protection against contaminants rather than to performing other essential missions. This makes the recipient of the NBC attack extremely vulnerable to follow-up conventional attacks. Using robotic vehicles will be one way to deal with the obstruction of repairs and maintenance, and to reduce delays and inefficiency imposed by NBC usage.

Obtaining attack assessments and other intelligence gaps may necessitate a reconnaissance of the contaminated area. Tactical requirements may also require combat vehicles to reach a strategic location by traversing across toxic terrain. Robots may furnish an edge on the toxic battlefield, and in the logistic infrastructure behind the front lines.

In addition to operation in hazardous areas, robots can also perform maneuvers too dangerous to be attempted with a man on board. Unmanned vehicles can execute maneuvers such as high-G turns and spins that would black out pilots. Design compromises made for human comfort and safety concerns can be eliminated. Robot weapon systems have greater stamina than do manned vehicles and do not degrade in performance over long periods of operation. Devoid of the human instinct of fear, robots can be risked in ways unthinkable for a human, including suicide missions. Donald A. Hicks, the former U.S. undersecretary of defense for research and engineering, has stated that he believes the United States cannot afford to risk so many lives and airplanes against the increasing lethality of Soviet defenses. He instead supports developing unmanned weapons. His belief no doubt is in part shaped by the December 1983 loss of three navy aircraft and one pilot to the Soviet-supplied Syrian surface-to-air missiles in the Bekaa Valley of Lebanon, which is juxtaposed to the success the Israelis had in using RPVs in the same Bekaa Valley seventeen months earlier.

Manpower shortages provide another reason for moving toward automated systems. Robots can take over less essential functions such as security, fire fighting, and some maintenance and logistic applications. Highly trained humans, such as skilled pilots, can be devoted to the most essential duties, reserving the robots to carry out less important or "unlikely to survive" missions.

MAJOR ROBOTIC PATHS

Robots are making inroads on the battlefield. Unmanned vehicle programs are proliferating throughout the world. The success achieved by the Israeli Scout and Mastiff RPVs in the 1982 Lebanon battles has spurred on RPV development in the United States, Great Britain, West Germany, Italy, South Africa, and the Soviet Union. The use of the French built PAP-104 ROVs by the British in the Falklands to clear mines, and the recovery of a new type Soviet mine by an ROV during the clearing operations in the Red Sea have highlighted robotic effectiveness in mine clearing. ROV programs exist now in the United States, Great Britain, France, West Germany, Sweden, and Italy. Ground robotic vehicle research that was pioneered in Great Britain as a means to combat terrorism is now being actively pursued in the United States and Israel. These mobile robots have already achieved acclaim in their handling and disarming of planted explosives, and one such robot rigged to carry a shotgun was actually used by police to shoot an intransigent armed criminal. The Japanese are working on advanced robotic wheeled, tracked, and legged vehicles and underwater systems. Although these are primarily designed for commercial use, the technology and some of the actual systems have direct military applications.

What Form Will They Take?

Upon accepting the premise that robots will emerge on the battlefield in significant numbers, the next question that is likely to arise is: What form or shape will they take? How combat robots will appear and act is largely dependent on whether they are teleoperated, autonomous, or some hybrid.

Within the defense scientific and engineering community a schism is developing between teleoperator researchers and the devotees of autonomous systems. Some teleoperator advocates believe that their preferred systems furnish the flexibility derived from "man-in-the-loop" operations.

They also view them as allowing for more and more programmable features to be incrementally added, which at some point will instill them with artificial intelligence, enabling them to operate autonomously, or through remote control. Other researchers claim that teleoperators and autonomous vehicles are based on dissimilar technologies, and the contrasting mind sets of their developers need different vehicle approaches. Trying to make their vehicles interchangeable lessens the optimal design and capabilities for each.

Examples of both views are reflected in different development programs. Odetics's Odex I and Robot Defense Systems's PROWLER are basically remote-controlled vehicles, which have some semi-autonomous capability through programmed instructions. Over time, more features are being added, such as obstacle avoidance, which will eventually lead to an autonomous system. On the other hand, the U.S. Navy's Robart and DARPA/Martin Marietta's Autonomous Land Vehicle are geared toward developing an autonomous system without any teleoperator antecedent. Technologically, it might be better to develop teleoperated and autonomous vehicles separately, although economic and political pressures may force the cohabitation of the two in the same vehicles and resources. The incremental switch from "man-in-the-loop" to autonomous in an evolutionary manner would be more acceptable to much of the military leadership. It is likely that military organizations will continue to emphasize teleoperated systems, while independent defense-oriented research organizations such as DARPA and the private sector will push autonomous technology developments.

Teleoperators will emerge first on the battlefield because their technologies are less complex and the military bureaucracies will be more comfortable with their existence. Greater payoffs will be derived from autonomous systems, whose development lies further down the road. Autonomous vehicles, however, pose some very important risks and questions for humanity's future, which will be addressed toward the end of this chapter.

The near-term adoption of robotic teleoperated vehicles has important implications for the kinds of "warriors" that military forces will retain, and the types of skills that must be recruited. If the command and control links between teleoperators and their human operators are made secure and free from disruption, and the signals can be transmitted over long distances, perhaps by satellites, then the nature of warfare and its warriors will change. A tank crew at the front could conceivably be replaced with a teleoperated vehicle controlled by someone in a Pentagon office. A teenager who possesses good hand-eye coordination and has quick reflexes (the kind of traits that allow one to perform admirably on video games) may be better able to undertake certain missions through teleoperated systems than could a young man in his prime at the scene of the battle. Without concerns for age, sex, and physical fitness, teleoperators offer the military a greater pool of people to select from, which may be important to Western countries experiencing the aging of their populations and greater competition from the private sector. Countries equipped with teleoperators may be more willing to engage in distant warfare.

Multi-environmental Robotic Vehicles

A major challenge in the field of unmanned systems is the development of multi-environmental robotic vehicles (MERVs). These are grown-up versions of the Japanese-made robot toys (Transformers, GoBots, etc.) that have so captured American youth. Not only can robotic vehicles adapt to more than one environment (land, air, sea, and space); in a limited sense, they already have, with the deployment of submarine-launched cruise missiles such as Tomahawk. These robot missiles are propelled through torpedo tubes into the water. When they break through the surface, some of the encasement is jettisoned, and fins appear, enabling the vehicle to travel through the air. More sophisticated MERVs will be a natural evolutionary trend for future robotics research.

One conceptual MERV system features a body with counter-rotating blades, similar in design to the Canadair CL-227 Sentinel, and is combined with a legged vehicle resembling an Odetics-type functionoid. This vehicle could travel over ground in dense foliage with the rotor blades in a folded position. It could also stay immobile for days in a concealed position while picking up valuable intelligence through its sensors. When activated, it could unfold its rotor blades, and convert into the helicopter mode for aerial surveillance and other airborne missions.

Another MERV approach involves an aerial remotely piloted vehicle that could dive into the water and continue on in a submerged mode as a remotely operated minisubmarine. This sort of system could carry out some nonacoustic submarine detection missions using a magnetic anomaly detection sensor or could tow an acoustic array while flying above the water. It could convert into an assault submersible, locking in on the enemy submarine for an attack. Not much work has been done in multi-environmental robotic vehicles. The field is ripe and the timing is right for some innovative companies to enter this forward edge of unmanned vehicle conceptual research.

Other Trends

As the close of the twentieth century nears, the following trends are influencing the first steps taken by combat robotics into air, sea, ground, and space environments.

Increased automation within major weapon systems, whether they be tanks, aircraft, or ships, has already resulted in a gradual reduction of personnel involved in their operations. Besides saving on costs associated with manpower and training, the replacement of crew members will reduce weight and space in manned vehicles. Automatic gun loaders have already cut tank crew size as well as reduced the overall size of the vehicle, thus increasing its survivability. The U.S. Air Force Pilots Associate Program and the U.S. Army Advanced Rotocraft Technology Integration (ARTI) Program are designed to eliminate the need for navigators, weapons officers, and other aircraft operator functions. These measures can be viewed as incremental steps of increasing automation. Slowly, more and more manned functions will be replaced by computers, electronic aids, and robotic appendages, until vehicles are completely automated, and any humans on board will be more passenger monitors than drivers and pilots.

Even more dramatic than this incremental approach is the direct move toward robotic weapon systems that is reflected in this book. The previous chapters have highlighted robotic systems currently in operational use, the numerous prototype vehicles and research efforts being undertaken, and a number of conceptual systems. Robots currently in development, or soon to be deployed, represent systems that for most part require less demanding technologies to perform fairly simple tasks, and that are not competing against manned systems. On the land, maintenance, logistics, and mine-clearing teleoperated robots are examples of vehicles facing the least resistance from manned advocates.

One type of robot that will make a near-term entry on the battlefield is the remotely controlled excavator. Air bases in Europe will be a likely target for precision munitions, including chemical weapons. Runways will be cratered and seeded with bomblets, which can stop aircraft operations. The first priority following an attack will be the filling of craters and removal of bomblets. Men in NBC suits will not be efficient enough to get the air base quickly back into operation. Through the use of remotely controlled Rapid Runway-Repair Excavators, the airbase can be efficiently repaired without risk to the human operators. Such a vehicle can be operated on board by a person or be remotely controlled by the addition of a "black box." Officials can determine when it is best to use humans to clear runways, or when the removal of bomblets and filling of craters might best be left up to robots. Mine-clearing robots that can traverse across booby traps or minefields without risking a driver are also candidates for a near-term field appearance.

The early combat robots will be small teleoperated vehicles equipped with light antitank missiles. Such systems can aid the West in offsetting Soviet armor's numerical superiority. They are less likely to be resisted by the manned system advocates because they will be viewed not so much as replacing tanks, but rather as eliminating the great danger to the overly exposed individual soldiers launching antitank missiles against armored vehicles.

Reconnaissance and decoy unmanned air vehicles will soon be joined by systems designed to perform offensive missions. Electronic warfare and munitions-laden RPVs will appear in significant numbers over the next few years. Assault RPVs will at first be of the expendable variety, although missile- and bomb-carrying systems will be common after the turn of the century. Many aerial unmanned systems will use fiber optic control links, although a number will also perform preprogrammed functions.

In the sea, mine-disposal ROVs will soon be complemented by tethered systems performing escort and submarine detection missions. They will not serve as replacements of manned craft but as adjuncts to their capabilities. Unmanned submersibles equipped with sophisticated manipulators will be employed to implant sensors, remove wreckage, and retrieve equipment. Shortly after these trends appear, navies will experiment with explosive-laden ROVs, followed by those that carry their own torpedoes, enabling sea-roving robots to have an offensive "punch."

The first space-based systems will be teleoperated work vehicles used to assemble manned space stations and the robot battle stations envisioned for SDI. Surprisingly, the largest amount of funding associated with combat robotics is directed towards the heavens. This is due to the massive budgetary and organizational support behind SDI.

The SDI concept is heavily dependent upon combat robotics technology. Unlike the "Star Wars" vehicles depicted in the movies, manned systems would be unable to operate at the efficiency rates needed in highly lethal space battle zones. The major option for SDI is unmanned weapon systems. The overall system is described by some researchers as being one giant, global distributive robot. This "mega-robot's" sensors would consist of early warning satellites and ground radar. The computer hardware and software of an advanced artificial intelligence architecture would serve as its processing brain, and its killers would be made up of antiballistic missile weaponry consisting of directed energy, kinetic penetrators, and other ballistic missile killers. The widely separated components of SDI must exchange data at such high rates of speed to counter a strategic attack that humans will be unable to participate as "on-the-spot" decision makers for the system's command and control, nor could they effectively monitor developments. The resulting system is therefore conceived as a giant autonomous robot dependent on the connectivity of its separate elements but acting independently of human involvement. The notion that such important decisions will be made without human involvement is greeted with great distrust by many individuals, including some renowned computer scientists, thus adding to this political debate.

Even if SDI falls short of being the overall comprehensive protective barrier for the whole country's populace as advocated by President Reagan, space-based robotic weaponry may still prove very useful for point defense roles of high-value targets such as missile silos, key command and control centers, and aircraft carriers. The artificial intelligence architecture required for point defenses is very complex, yet it is much more feasible than a comprehensive SDI because it does not require the interconnectivity between so many sensors and weapons platforms. The warfare environment of space-based battle stations equipped with nuclear-pumped X-ray lasers, electromagnetic rail guns, and charged particle beams is far too toxic for humans and any space battle stations are likely to be robotic rather than manned.

AUTOMATED TERRORISTS AND COUNTERTERRORISTS

A quasi-military spin-off for robotic applications is in the unconventional warfare area of terrorism. Unmanned systems can be used by terrorists and as a means to protect civilian populations from this menace.

To blow up people indiscriminately, shoot into crowds, and to perform other terrorist acts requires individuals with psychopathic personalities. Fanatical and revolutionary organizations recruit such people to further their causes. There is a growing concern that robotic vehicles with no moral conscience, and without any fear of suicide missions, might simulate the same psychopathic character traits and therefore make ideal terrorists. The use of mechanized killers would certainly cause panic and concern among victims and generate the publicity sought by terrorists. Libya's purchase of the Italian Meteor firm's Excalibur remotely controlled boats and their shipment to radical groups indicate that terrorists have realized the potential that robotics offers to the spread of murder and mayhem. Automated trucks loaded with TNT, submersibles packed with explosives, and remotely piloted aerial bombs could also pose serious problems if they wind up in terrorist hands. Defense, law enforcement, and intelligence organizations would be wise to monitor and control the flow of unmanned systems and technology.

Although robots can perform terrorist acts, either through programmed instructions or remote control, they can also be used to combat terrorism. Robotic vehicles will not eliminate terrorism, but teleoperated and autonomous air, land, and sea vehicles can patrol airports, embassies, harbors, and other potential target areas. Patrol and surveillance functions may be performed less expensively, less obtrusively, and less dangerously than with manned systems. By using advanced automation and robotic technology, it may be possible to stop and inspect vehicles suspected of carrying bombs without direct risk to U.S. personnel. Their use might prevent tragic events such as the 1983 suicide truck bombings of the U.S. Marine Corps headquarters and U.S. embassy buildings in Lebanon. Robots Defense Systems's PROWLER is an example of an unmanned vehicle capable of patrolling sensitive facilities and performing activities that could stop and inspect incoming vehicles.

Remotely piloted vehicles using advanced cameras and sensors (including infrared, acoustic, and radar) have shown great promise in performing discreet surveillance missions and in jamming terrorist communications and targeting them for smart weapons. The Israelis have used an RPV to follow a car being driven evasively by terrorists. After several hours, the RPV tracked the terrorists to their headquarters, where the Israelis destroyed them with a conventional air attack. Remotely piloted blimps such as the RPMB built by Developmental Sciences, Inc. and inflatable RPVs manufactured by ILC Dover and Stewkie Aerodynamics can conduct surveillance operations in cluttered urban areas. Such vehicles could hover in the same position for days at a time, with their cameras and sensors monitoring target areas and selected activities from overhead.

The disposal of terrorist bombs requires individuals who exhibit nerves of steel and great courage. Although many servicemen have these character traits, explosive ordnance disposal can also be performed by robotic vehicles. Human lives only have to be risked when the bombs are inaccessible to the unmanned system. The British perfected the use of robotic vehicles in the 1970s to cope with the Irish Republican Army terrorist bombings in Northern Ireland. Prior to the introduction of EOD robots, an explosives expert would attempt to defuse bombs by removing the bomb's components by hand. This would usually involve lifting off the cover of the device and then disabling the detonator. The bomb removal specialist would either delicately try to disconnect the detonator, or would attempt to destroy the bomb in a controlled explosion. Many fingers, arms, and even lives have been lost due to some slight error. Use of a robot allows the bomb disposal experts to stay a safe distance away from the explosives. The robot is directed towards the bomb and transmits pictures to the operator through its T.V. camera. The operator can then instruct the robot to lift parts of the explosive device with its manipulators. It can either disarm the bomb or destroy the mechanism with a shotgun blast. Unfortunately, terrorists are becoming more sophisticated in countering robotic technology. There was one instance in which revolutionaries were able to override the EOD operator's radio control and have the robot turn on him. The operator barely escaped being blown up by his own robot. This problem was eliminated by switching to a more secure tether-controlled robot.

ROBOT POLITICS

Some defense analysts are skeptical of the importance that high-technology weaponry has on the battlefield and they are likely to have this same attitude on combat robotics. They contend that the "gee-whiz" technology of U.S. forces in Vietnam was ineffectual in combatting the Vietcong and North Vietnamese and that the Nazi superiority in advanced weaponry over the Allies could not compensate for their inferiority in numbers and the massive weapon-producing capability of U.S. industry.[1] Some members of the military reform movement, including those within the U.S. Congress, have also expressed dissatisfaction with America's infatuation with "high-tech" weaponry as a cure for all of our national security ills. Former U.S. Senator Gary Hart and William Lind point out in their book, *America Can Win*, however, that reformers do not want quantity instead of quality. Rather, they seek weapons that in general are smaller, simpler, and less expensive.[2] Many robotic systems, including RPVs, can certainly fill this bill. Superior weaponry without good strategy and tactics is a formula for defeat. As Colonel Trevor Dupuy so adeptly demonstrates in his book, *The Evolution of Weapons and Warfare*, technology is only one factor among many that determine the victor in any engagement.[3] Yet it is a very important factor that can provide that extra edge in winning battles and wars. Clearly, robotic weapons are not a panacea for all defense needs, yet they will play an essential role in strategic and conventional defense over the next decade, and on into the twenty-first century.

Opposition to Robotics Use

Within the military bureaucracies, political infighting exists between the manned and unmanned advocates for control over budgetary resources. The conflict, however, is not only concerned with dollars but also about how military personnel who man weapon systems view their importance to the bureaucracy they serve and how a number of them feel threatened by robotic systems that may take some of their roles. The resistance of factory workers to the entry of industrial robotics is minor compared to the unwillingness that some of the military has toward accepting combat robotics. In the factory it is the blue-collar workers whose jobs face extinction by automation, and it is the upper management or white-collar workers who stay employed and even prosper from the greater efficiency furnished by robots. In the military, however, oftentimes the upper management are hands-on operators of weapon systems; many of their roles are at stake because of the potential introduction of robotic vehicles. Their resistance is likely to be fiercer than what has occurred in the factories.

The military bureaucracies throughout the world are primarily run by the skilled operators who have been promoted up through the ranks. In the United States, the air force flag rank is largely made up of pilots. In the navy, both aviators and ship commanders are in control of the organization. In today's army, command accrues primarily to those associated with combat soldiers. It is the same in other nations' military establishments. Planners, intelligence officers, communications officers, acquisition managers, and other noncombat specialists rarely reach the pinnacle of power. Therefore, there is natural reluctance on the part of the leadership of military bureaucracies to embrace robotic vehicles that may reduce the influence of the manned systems and their operators. One approach to appease the manned vehicle proponents is to develop vehicles that can be operated on-board by a person or be remotely controlled by the addition of a "black box." John Deere's Rapid Runway-Repair Excavator is an example of such a system, which can operate manually or through automation. Ultimately, factors such as high combat attrition, manpower problems, high material costs, escalating personnel costs, and the deterioration of the defense industrial base will ultimately erode the military bureaucracy's ability to resist unmanned systems.

Individuals less directly involved in the weapons acquisition process than the military reform caucus of the Congress and the Department of Defense bureaucracy are concerned about robotics and will increasingly become part of the American defense debate. A weapons system or concept that raises societal concerns of ethics and morality, galvanizes a cadre of scientists and politicians, and even mobilizes grass roots support for or against its development certainly indicates that it is of great importance to society. Of all the weapons programs initiated each year, relatively few generate such widespread political attention and emotion. President Reagan's SDI and the basing of the MX missile during the Carter presidency are examples of weapons programs whose merits were debated at the societal level.

Robotic weapons systems have reached a level of technological maturity and are no longer considered figments of science fiction writers' imaginations. The general scientific community, as well as political activists, now visualize their deployment and widespread use on the battlefield. The importance of unmanned weapons has led to the formation of robotics interest groups, such as the 1,100-member Computer Professionals for Social Responsibility, headquartered in Palo Alto, California. Gary Chapman, one of the organization's spokespersons, has labeled unmanned weapon systems "killer robots" and proclaimed their development to be illegal and one that will bring about the next scientific ethics crisis. Chapman believes that combat robotics are illegal under the Geneva Convention's rules of war because a robot would be unable to recognize a surrendering person; under the convention, it is illegal to wage warfare in a way that does not allow the other side to surrender. Chapman maintains that a soldier who randomly shoots in an area, killing everyone, can be placed on trial for war crimes. He considers robotic weapons to be such indiscriminant killers, comparable to biological and chemical agents that are condemned by the World Court and the United Nations.[4] Retired General Tralford Taylor, chief U.S. prosecutor at the Nuremburg war crimes trials, has advised Chapman to take his case against unmanned weapon systems to the World Court. Massachusetts Institute of Technology professor Dr. Joseph Weizenbaum, who pioneered the development of artificial intelligence, also shares such sentiment and finds the concept of robot warriors to be abhorent, although he has little doubt that military forces will eventually field autonomous robots.[5]

Proponents' Arguments

Several arguments have been made to refute Chapman's concerns about robotic weapons systems. Unmanned weapons systems are primarily being developed to destroy enemy systems, including tanks, aircraft, and other "high-value" targets. The munitions they carry are not designed to pick and choose their victims among individual foot soldiers. People will certainly be killed, perhaps including innocent civilians. Manually operated artillery, bombs dropped by manned aircraft, and rounds shot from tanks also bring about similar destruction and loss of life. Unmanned systems of today's technology would be unable to recognize an individual located in the midst of an important target with his arms raised in surrender. The same, however, is true of major manned systems, which are dependent on many of the same sensors and technology that robots use for targeting.

Teleoperated robotic vehicles respond to the decisions made by a human operator, who is receiving real-time visual, audio, and other sensory data that simulate the "awareness" of actually being on board the vehicle, without exposing the human to danger. If an operator were able to view a surrendering enemy soldier through the camera system he/she might hold his/her fire, as would a gunner in a manned vehicle. Autonomous combat robotic systems currently under development are designed to attack high-value targets on the battlefield and would ignore individuals not in vehicles or installations. Autonomous security robotic systems such as Robart II are being built to detect and incapacitate intruders, but they are also being programmed to recognize surrender and to use nonlethal force. The use of robotic weapon systems may be no less humane than manned weapon systems and there is no reason that they must violate established codes of warfare.

Several scientific and engineering organizations espouse the development of artificial intelligence and military robotics. The 1,800-member Association for Unmanned Vehicle Systems (AUVS), whose national headquarters is located in Washington, DC, has focused its activities primarily on influencing robotic weapons development within the U.S. Department of Defense. It has, however, made several efforts to influence the larger society through videotape presentations to the mass media and through its quarterly publication, *Unmanned Systems*.

The questions raised by Chapman are important ones, and it is healthy for a democracy to subject governmental programs to moral and legal questions, as well as to technological and operational analysis. The activities of groups such as Computer Professionals for Social Responsibility and the Association for Unmanned Vehicle Systems are essential for a vital democracy to function properly by bringing to the public and policymakers salient points on important defense issues. The debate about robot weapon systems will intensify in the future, and has been recently stimulated by a major U.S. program, President Reagan's SDI.

The American political mainstream in both the Republican and Democratic parties seems to accept the need for a strong national defense. Advocates of SDI view it as a means of removing the nuclear "sword of Damocles" posed by the doctrine of Mutually Assured Destruction (MAD). Some SDI supporters perceive SDI as a way to survive and win a nuclear war, whereas others maintain that it will serve as a greater deterrent to keep Soviet actions and motives in check. Opponents claim that the system is unworkable, destabilizing, and takes away resources that should be going to conventional forces. Donald A. Hicks has warned that research money pumped into SDI has resulted in a reduction of the opportunities for technological breakthroughs in other areas of the kind on which the United States has relied to compensate for Soviet numerical superiority.[6]

ROBOTS RUNNING WILD?

Teleoperated vehicles will gain a foothold onto the battlefield during the eve of this century. By the dawn of the twenty-first century, they will be firmly entrenched. Their reign, however, will be short-lived, lasting only several decades. They will be gradually replaced by hybrid systems that are basically teleoperated, but that also have a limited ability to perform certain functions without direct human control. The shift will quickly gain in momentum as the technology matures enough to enable systems instilled with "artificial intelligence" to effectively undertake combat missions. By the middle of the twenty-first century, autonomous vehicles will be preeminent on the battlefield, replacing most unmanned as well as manned systems. The trend toward autonomous vehicles also carries with it major concerns that anyone interested in this subject must be alert to. By releasing this "genie out of the bottle," humanity's survival may be in jeopardy.

Many critics of unmanned weapon systems are wary of autonomous robots making important decisions with their human-manufactured intelligence. V. Daniel Hunt, author of several books on industrial robots and artificial intelligence, has asked,

> Can we allow an autonomous battlefield creation to roam the battlefield searching out the enemy who is defined by a software algorithm? What are the implications and what cost are we willing to incur for a wrong or errant machine decision made by this hardware? Has it been calculated? The escalation of large weapon systems run by internally controlled elements is inevitable; so also is the escalation of effect to be felt should that system revert to an unpredicted operational mode.[7]

As with all mechanisms, combat robotic vehicles will have their share of breakdowns and faulty operations. Programming bugs will never be completely eradicated, although a great deal of attention to testing and quality control can limit software problems. The best that can be achieved is that the difficulties posed by malfunctioning robots will be controllable. The same, however, is true of manned weapon systems. Humans periodically exhibit irrational behavior and pose as great a risk as do renegade robots. For several years we have placed nuclear warheads "in the hands" of robotic cruise missiles. If they malfunction, the consequences can be disastrous. A bomber or missile crew that cracks under pressure or allows emotion to dictate actions may also cause a calamity. In the long run, the benefits derived from robotic use will outweigh the risks. Robots conduct missions without fear and if captured, a self-destruct device could be activated. At least we can be sure that in the near term robots will not be sophisticated enough to betray their country for financial gains, sexual favors, or ideological reasons.

On the other hand, the long-term implications of autonomous vehicles are difficult to assess. Some of the better scenarios on the future of "smart" military robots have come from the creative minds of science fiction writers rather than from military futurists. If it were not for the robot stories of Capek, Asimov, and other such writers, the field of robotics would not be as advanced as it is today. These literary visionaries have inspired numerous scientists to enter the fields of robotics, artificial intelligence, and sensor research, and have often set the goals and milestones to gauge actual scientific progress.

Keith Laumer, a noted science fiction writer, has outlined a fictional evolution of unmanned tanks, which is as sound as any forecast that a robotic scientist could project. His Bolo series of vehicles begins development in General Motor's Bolo division in 1989. The initial vehicle, the Bolo Mark I Model B, is an enlarged (150-ton/136-metric ton), enhanced version of a main battle tank. It has powered assisted servomechanisms. In battle, it is normally operated by a three man crew; however, it also has the capability to perform a number of operations in an unmanned mode through preprogrammed instruction. Patrol duty is one function that can be performed with no crew onboard. The follow-on model appears in 1995 and is christened the Mark II. This tank is more highly automated than the Mark I and has a sophisticated on board fire control computer that eliminates the need for two crew members. Only one man is needed to operate the vehicle in battle. A successor vehicle has greater fire power and endurance, but takes a step backward by requiring a crew of three humans to handle its intricate mechanisms. Being somewhat conservative, Laumer does not describe a completely automated Bolo until the introduction of the Mark XV in the twenty-fifth century. This vehicle no longer carries a person as an operator but only as a passenger. The first self-aware Bolo, the Mark XX model, is finally introduced after much resistance from high-level military officials, who fear its running amok and wreaking havoc on its owners. By the time of the Bolo Mark XXX, human tactical and strategic thinking no longer plays a part in combat, which has been supplanted by the conceptualization and decision-making abilities of the robotic vehicles.[8]

The developments described in this book may be laying the groundwork for long-term results similar to Laumer's scenario. There are, however, other aspects to autonomous unmanned vehicles that should be addressed. Science fiction works from Karel Capek's 1917 short story *Rossum's Universal Robots* on through contemporary movies such as *2001: A Space Odyssey* and the *Forbin Project* warn that artificial intelligence in robots will some day lead to humanity's downfall. Asimov's Three Laws of Robotics described in Chapter 1 were designed in part to thwart any robotic ability to overthrow humanity. Of course, the adoption of such laws would eliminate combat use of robots; thus they would not be adopted. But the dangers are real and future robotic weapons developers must determine what level of intelligence is safe in their robots. With the mega data storage and high processing speeds of tomorrow's computer "brains," robots would be able to solve equations, deduce the answers to certain problems, and react to some circumstances much quicker than could humans. Robots will most likely be equipped with tactical and perhaps some strategic decision-making capabilities. Allowing them to make major strategic and doctrinal policy decisions could lead to the supplanting of humanity's control. Our successors must carefully decide where to draw the line. Future scientists must also decide whether it is best to implant emotions or to make robots cold, detached, and highly logical thinkers. Either direction has its share of positive and negative features.

When developing artificial intelligence for robotic vehicles, it is quite natural for engineers and program managers to relate it to human intelligence, and to try to mimic human thinking. Patterning the robot's behavior after nonhuman types of intelligence or the instinctual behavioral traits exhibited by lower forms of life may be more appropriate for certain military missions. Programming a dune-buggy type robot equipped with missiles to attack enemy tanks or an unmanned assault submersible to go after enemy submarines may best be achieved by developing systems that resemble the behavioral processes in which soldier ants, wasps, or piranha attack and inflict wounds on other creatures.

The SHARC program mentioned in Chapter 5, in which DARPA is examining the collective interaction and cooperative behavior among a group of autonomous robotic vehicles, may produce a type of intelligence that incorporates telepathic understanding and communication. This would be based on continuous data links among the vehicles. At some future date, an enhanced form of parallel data processing would result from each vehicle performing a set of computations, which feeds into the overall data network of the other autonomous vehicles. The sum total of intelligence would be much greater than any of the individual pieces. Future robotic vehicles may use modes of reasoning and perception that may not resemble anything that exists in the natural world. A very great danger lies in the development of artificial creatures whose thought patterns and motivations are not clearly understood.

Man is currently the master. Man is moving toward a battlefield without humans. Man will certainly give robots more and more intelligence and data processing capability, and man must be increasingly concerned that these capabilities remain under control. Knowledgeable citizens, elected officials, decision makers, and military leaders must prepare now for the revolutions that robots will cause as they enter the future battlefield. Thinking must adjust to the rapid pace of change to ensure that our society and civilization are not caught unawares as other civilizations were by the invention of the longbow, gunpowder, and other military technologies. It is imperative that we understand the revolutions that will occur in the lethality of weapons, leadership styles, and personnel and organizational requirements. Ultimately, as many of the robotic developments we have cataloged turn into weapons systems, we must all be sure that these robots are no less moral than those humans who design and use them.

NOTES

1. Andrews, R. "The Illusion of Conventional Defenses." *National Security Record*, September 1986, p. 5. Andrews provides an in-depth discussion of the limitations of high technology for conventional defense.
2. Hart, G. and Lind, W. *America Can Win: The Case For Military Reform*. Adler & Adler, Bethesda, MD, 1986. Hart and Lind furnish the reader with the reform movement's perspective towards high technology and its use on the battlefield.
3. Dupuy, T. *The Evolution of Weapons and Warfare*. Bobbs-Merrill Co., Indianapolis/New York, 1980. Dupuy provides an excellent historical survey of the interrelationship between technology and warfare.
4. Beers, D. "Killer Robots Bring Problems: Experts Call Such Weapons Next Crisis In Scientific Ethics." *Cleveland Plain Dealer*, September 28, 1986, p. 28.

Beers describes the rationale and effort against robotics system development being waged by Chapman and the Computer Professionals for Social Responsibility.

5. Beers, "Killer Robots." Weizenbaum's sentiments are detailed in the Beers article.
6. Wilson, G. "Budget Cuts: SDI's Priority Hurts Conventional Military Readiness." *Washington Post,* October 21, 1986, p. 10. This article highlights the arguments that SDI is harming conventional forces through slashed budgets in readiness and research and development.
7. Hunt, V. Daniel. "Should Machines Make The Decisions?" *Washington Technology,* December 11, 1986, p. 10.
8. Laumer, K. *Rogue Bolo*. Baen Books, New York, 1986. This book contained Laumer's fictional history of the evolution of autonomous robotic combat vehicles.

Bibliography

"AAI Corp. Will Deliver Initial Pioneer I to Navy in May." (1986) *Aviation Week & Space Technology,* April 28, 109.

"A Better Blimp." (1982) *OMNI*, September, 46.

Abronson, R. (1984) "Robots Go To War." *Machine Design*, December 6, 72–79.

Adams, R., Lehman, L. and Herskovitz, M. (1985) "A Ship-Based High Altitude RPV Study." *Unmanned Systems*, Winter, 25–29.

Advanced Technology Advisory Committee, National Aeronautics and Space Administration. (1985) *Advancing Automation and Robotics Technology for the Space Station and For the U.S. Economy*. (Vol. 1—Executive Overview). Washington, DC: Author.

Advanced Technology Advisory Committee, National Aeronautics and Space Administration. (1985) *Advancing Automation and Robotics Technology for the Space Station and For the U.S. Economy* (Vol. 2—Technical Report). Washington, DC: Author.

"Aerodyne Plans to Enter CH-84 in Army, Navy RPV Competitions." (1986) *Aviation Week & Space Technology,* April 28, 108.

"AFIT Seeks Tactical Aid Based on AI." (1986) *Aviation Week & Space Technology,* February 17, 61–65.

"Air Force Developing Smart Lethal Expendable." (1983) *Journal of Electronic Defense,* July, 59–60.

"Air Force to Test 'Brilliant Bomb' That Seeks Out Targets 10 Miles Away." (1986) *NYC Tribune,* June 18, 5.

"Air Force Will Begin Testing on New Guided Bomb." (1986) *Defense News*, July 7, 28.

Alterman, S. and Stolfo, S. (1985) "The Application of Parallel Processor Technology to Future Unmanned Vehicle Systems." *Unmanned Systems,* Fall, 10–19.

Andreone, V. (1982) "Aldebaran Remote Boat and Tracking System." *Unmanned Systems,* Fall, 18–25.

Andrews, R. (1986) "The Illusion of Conventional Defenses." *National Security Record,* September, 5.

"Aquila/RPV Program." (1985) *Defense Department Authorization and Oversight Hearings on H.R. 1872,* H.A.S.C. No. 99–2, 304–318.

"Army Completing Development Tests For Lockheed Aquila RPV System." (1986) *Aviation Week & Space Technology,* April 28, 85–89.

"Army Develops High-Altitude Offset Airdrop System." (1986) *Aviation Week & Space Technology,* June 16, 77.

Aronson, R. (1986) "Weight Is The Enemy in New Tank Design." *Machine Design,* April 24, 32.

Asimov, I. (1968) "The Perfect Machine." *Science Journal,* October 1968, 115–118.
Asimov, I. and Frenkel, K. (1985) *Robots: Machines in Man's Image.* New York: Harmony Books.
Association of Unmanned Vehicle Systems. (1985) *Right on Target Proceedings,* July, Anaheim, CA: Author.
"Aussies, Navy Near Agreement On Ship Missile Decoy." (1986) *Navy News & Undersea Technology,* April 11, 6.
"Australia's Remote-Control Parachute System." (1985) *International Defense Review,* September, 1516.
Automation & Robotics Panel, California Space Institute, University of California. (1985) *Automation & Robotics For The National Space Program.* California: Author.
"The Autonomous Remotely Controlled Submersible (ARCS)." (1985) *SubNotes,* June, 14–15.
Banks, T. (1985) "Experience in Lebanon Enhances Mazlat's Chance at RPV Contract." *Defense News,* October 21, 8.
Barnaby, F. (1984) *Space Weapons.* New York: Gallery Books.
Barrett, F. (1985) "The Robot Revolution." *The Futurist,* October, 37–40.
Bartholet, T. (1986) "Odetics Pioneering Development of Advanced Intelligent Machines." *Unmanned Systems,* Spring, 13–15, 30.
"Beating Swords Into Lemons." (1984, September 27) *The New York Times*, p. 18.
Bedard, P. (1985) "Drones to Ride Shotgun For Ships." *Navy News & Undersea Technology*, September 27, 1, 7.
Beers, D. (1986, September 28) "Killer Robots Bring Problems: Experts Call Such Weapons Next Crisis in Scientific Ethics." *Cleveland Plain Dealer,* p. 28.
Beggs, J. (1984) "Space Station: The Next Logical Step." *Aerospace America,* September, 47–52.
Beyers, D. (1986) "AF Targets Guided Bomb as First Brilliant Weapon." *Defense News,* December 29, 1, 21.
Bode, B. (1984) "Modifications of A Commercial Excavator for Rapid Runway Repair." Paper presented at the Government/Industry & Exposition, May, Washington, DC (SAE Technical Paper Series).
"Boeing Electronics Rolls Out Experimental RPV." (1986) *Aviation Week & Space Technology,* April 7, 29.
Bond, N. (1983) "Japanese Progress in Robotics: Tokyo Conference and Exhibition." *Scientific Bulletin Department of the Navy Office of Naval Research Far East,* October–December, 77–110.
Brady, E. and Finkelstein, B. (1982) "Unmanned Weapon Systems: A Nuclear Deterrent?" *Unmanned Systems*, September, 10–13, 27.
"Brandbury Refines Composite RPV Airframes for Upcoming Coast Guard, Army Competition." (1986) *Aviation Week & Space Technology,* April 28, 127–128.
"British Army Selects Real-Time Remote Artillery Direction System." (1986) *Aviation Week & Space Technology,* April 28, 61.
"British Companies Develop Range of Surveillance, Target Systems." (1986) *Aviation Week & Space Technology*, April 28, 91–97.
"British RPV With Infrared Sensor Will Produce Real-Time Data." (1985) *Aviation Week & Space Technology*, April 8, 42.
Britton, P. (1984) "Engineering The New Breed of Walking Machine." *Popular Science*, September, 66–69.
Broad, W. (1986, February 16) "Submersible Joins in Search for Challenger Debris." *The New York Times*, p. 1.

Burgess, R. (1973) *Ships Beneath the Sea: A History of Subs and Submersibles.* New York: McGraw-Hill.
Busby, F. (1985) "ROVs: Uncertain Times." *Sea Technology,* March, 41.
"Cable-Controlled Underwater Recovery Vehicle (CURV III)." (1982) *Seahorse,* Spring, 10.
Calder, N. (1968) *Unless Peace Comes: A Scientific Forecast of New Weapons.* New York: Viking Press.
Caldwell, H. and Kennedy, F. (1982) "RPV—Stepchild of Unmanned Vehicles." *National Defense,* September, 16–20.
Capek, K. (1973) *Rossum's Universal Robots* (P. Selver, Trans.) New York: Washington Press (original work published 1921).
Carlisle, R. (1984) "Space Station: Technology Development." *Aerospace America,* September, 60–66.
Carus, W. (1984) "U.S. Procurement of Israeli Defense Goods and Services." *AIPAC Papers on U.S.–Israeli Relations.* Washington, DC: American Israeli Public Affairs Committee.
Cerny, J. (1986) "Land Robot Revolution." *Unmanned Systems,* Spring, 12–13, 30.
Chastain, A. (1981) "Unmanned Aerial Vehicles." *Military Intelligence,* July–September, 32–36.
"CL-289 Reconnaissance Drone." (1983) *Armed Forces Journal,* December, 51.
Clark, R. (1977) *The Role of the Bomber.* New York: T.Y. Crowell Co.
Corddry, C. (1986, June 29) "Senate Panel Urges Shift in SDI Goals: Defense of Military Sites Emphasized." *Baltimore Sun,* p. 1.
Cornett, P. and Morrison, A. (1983) "Part of the Battlefield of the Future." *National Defense,* October, 24–30.
Covault, C. (1985) "Shuttle Mission EVAs to Demonstrate Space Station Assembly Techniques." *Aviation Week & Space Technology,* November, 63–69.
Cushman, J. (1985) "Undersea Robots For Mines." *Navy News & Undersea Technology*, March 29, 5.
Craig, T. (1984) "Remarks on Aerial Targets." *Unmanned Systems,* Winter, 26–27.
"DARPA's Pilot's Association Program Provides Development Challenges." (1986) *Aviation Week & Space Technology,* February 17, 45–52.
"DARPA Seeks Mobile Battlefield Robot Capable of Thinking for Itself." (1984) *Military/Space Electronics Design*, December, 10–11.
Davies, O. (Ed.). (1984) *The OMNI Book of Computers & Robots.* New York: Zebra Books.
Deken, J. (1986) *Silico Sapiens: The Fundamentals and Future of Robots.* New York: Bantam Books.
DePauk, P. and Wullert, J. (1985) "Navy Technology Requirements for Unmanned Airborne Vehicles." *Unmanned Systems,* Fall, 38–40.
"Development Sciences Prepares Skyeye for Army Competition." (1986) *Aviation Week & Space Technology,* April 28, 68–83.
"Dornier Developing Argus 2 Battlefield Surveillance System." (1985) *Aviation Week & Space Technology,* June 10, 73.
DrevDahl, C. (1983) "Mobile Robotics Will Serve Many Roles in Future Land Warfare." *Defense Systems Review,* November, 13–17.
Dupuy, T. (1980) *The Evolution of Weapons and Warfare.* Indianapolis: Bobbs-Merrill Co.
Ebisch, R. (1984) "Metal Warriors." *The Ambassador,* October, 22–30.
"ECA Develops Epaulard Unmanned Submersible." (1985) *Jane's Defence Weekly,* April 20, 68.

Elson, B. (1980) "Mini-RPV Being Developed For Army." *Aviation Week & Space Technology,* January 7, 2–7.

Esienstadt, S. (1986) "Lear Siegler and Egypt Near Agreement on Skyeye RPVs." *Defense News,* February 3, 1.

Everett, H. (1985) "Robotics In The Navy, Part II: Nonindustrial Development Efforts." *Robotics Age,* December, 9–12.

Everett, H. (1985) "Shipboard Applications—Are The Robots Really Coming?" Paper presented at the 22nd Annual Technical Symposium, Association of Scientists and Engineers of the Naval Sea Systems Command, Washington, DC.

Falconer, N. (1985) "Unmanned Vehicles and the Future of Air Power." *Jane's Defence Weekly,* April 20, 678–681.

Falconer, N. (1984) "Unmanned Vehicles and Their Impact on Defense." *Unmanned Systems,* Fall, 36–40.

Fialka, J. (1986, April 30) "Army Engineer's Anti-Tank Missile Proves Cheap, Reliable and Tough to Sell at Pentagon." *The Wall Street Journal,* p. 64.

Fialka, J. (1984, November 23) "Simple Army Drone Grows Complicated, Expensive and Late." *The Wall Street Journal,* pp. 1, 6.

Finkelstein, R. (1986) "A 6000 Ship Navy?" *Unmanned Systems,* Spring, 7–8.

Finkelstein, R. (1985) "Terrorism, Ants and Unmanned Vehicles." *Unmanned Systems,* Summer, 40, 45.

Finkelstein, R. (1983) "TROIKA: A Concept for a Robotic Weapon System." *Unmanned Systems,* Fall, 6–7.

Finnegan, J. (1985) *Military Intelligence: A Picture History.* Arlington, VA: U.S. Army Intelligence and Security Command.

Fitzsimons, B. (Ed.). (1978) *The Illustrated Encyclopedia of 20th Century Weapons and Warfare.* New York: Columbia House.

Flynn, A. (1985) "Redundant Sensors for Mobile Robot Navigation." Thesis for the degree of Master of Science at the Massachusetts Institute of Technology.

Ford, B. (1969) *German Secret Weapons Blueprint For Mars.* New York: Ballantine Books.

"Four-Footed Friend." (1969) *Newsweek*, April 14, 80.

Frederick, R., Rappolt, F., Pogust, F. and Kennedy, F. (1985) "Considerations for Sea-Basing Unmanned Aircraft for Tactical Support." *Unmanned Systems,* Spring, 31–38.

Freitag, R., Lottman, R. and Wigbels, L. (1984) "Space Station: The World Connection." *Aerospace America,* September, 76–80.

"French Continue Conservative Pace of Development Programs." (1986) *Aviation Week & Space Technology,* April 28, 113–116.

Friedrich, O., Branegan, J. and Hannifin, J. (1986) "Looking for What Went Wrong: NASA Begins an Arduous Search for Clues to the Disaster." *Time,* February 10, 36–37.

Fulghum, D. (1986) "Lockheed's Aquila RPV Flies Into Congressional Budget Buzzsaw." *Defense News,* July 21, 6–7.

Fulsang, E. (1985) "AI and Autonomous Military Robots." *Unmanned Systems*, Spring, 8–16.

Fulsang, E. (1985) "Robots On The Battlefield." *Defense Electronics,* October, 77–82.

Gallagher, E. (1970) *A Thousand Thoughts on Technology and Human Values.* Bethlehem, PA: Humanities Perspective on Technology Program, Lehigh University.

Garvin, F. (1985) "Robots Go To War." *International Combat Arms,* July, 14–23.

Gaudiano, M. (1984) "Electronic Commandos." *Journal of Electronic Defense*, April, 81–82, 89.
"GEC Avionics Given Phoenix Lead." (1985) *Flight International*, March 2, 11.
"GEC Avionics Wins Phoenix Contest." (1985) *Military Technology*, April, 104.
Gilmartin, T. (1986) "Boeing Wins Rocket Booster Contract." *Defense News*, February 24, 3.
Gilmartin, T. (1985) "NASA Readies Orbital Maneuvering Craft For Fast Production Takeoff By June." *Defense News*, October 21, 11.
Gilmartin, T. (1986) "NASA Taps TRW to Develop Tug For Space Shuttle Fleet." *Defense News*, June 30, 1.
Gossett, T. and Velligan, F. (1982) "The Aquila: A Versatile, Cost-Effective Military Tool Shows Its Potential." *Military Electronics/Countermeasures*, December, 74–78.
Greeley, B. (1985) "Navy RPV Program Receives Bids From Only Two Firms." *Aviation Week & Space Technology*, October 7, 16–17.
Greenwald, J. and Van Voorst, B. (1985) "Over Hill, Over Dale . . . Tomorrow's Weapons Are Being Designed Today." *Time*, August 19, 18–19.
Gregory, W. (1986) "Canada Adapting Shuttle Remote Arm To Space Station Service Facility." *Aviation Week & Space Technology*, March 3, 73–74.
Grier, P. (1986, May 29) "Mini-Missile, Major Questions." *Christian Science Monitor*, p.1.
Hambley, C. (1987) "Remote Possibilities." *Sea Power*, January, 24–35.
Harms, G. (1972) "Aerodyne." *Dornier-Post*, February, 19–20.
Harrison, H. (1962) *War With The Robots*. New York: Pyramid Books.
Hart, G. and Lind, W. (1986) *America Can Win: The Case for Military Reform*. Bethesda, MD: Adler & Adler.
Hartsfield, J. (1974) "A New Challenge." *National Defense*, March–April, 428–430.
Harvey, D. (1986) "Flying Today's Airplanes Into Tomorrow's Navy." *Armed Forces Journal*, January, 54.
Harvey, G. (1985) "Plans For Pilotless Planes." *Navy News & Undersea Technology*, August 2, 9.
Healy, M. (1985) "U.S.–Israeli Deal: Lehman Lays Groundwork on Drones, Boats, Subs." *Navy News & Undersea Technology*, April 26, 10.
Hiatt, F. (1986, June 27). "Space Launch Needs of SDI Are Estimated: 600 Liftoffs Seen in 3-Year Period." *The Washington Post*, p. A24.
Hiatt, F. (1986, February 6) "Space Plane Soars on Reagan's Support." *The Washington Post*, p. A3.
Hilts, P. (1986, July 16) "Robot Video Camera Starts In-Depth Tour of the Titanic." *The Washington Post*, pp. A1, A6.
Hilt, P. (1986, July 17) "Explorers Marvel At Titanic's Glass." *The Washington Post*, p. A4.
Hirose, S. (1984) "A Study of Design and Control of a Quadruped Walking Vehicle." *The International Journal of Robotics Research*, Summer, 113–133.
Hirose, S. and Umetani, Y. (1976) "Kinematic Control of Active Cord Mechanism With Tactile Sensors." In *Theory and Practice of Robots and Manipulators, Proceedings, Second International CISM–IFToMM Symposium*, Warsaw, Poland, September, 241–251.
Hirose, S. and Umetari, Y. (1978) "Some Considerations On A Feasible Walking Mechanism As A Terrain Vehicle." Paper presented at the Third CISM–IFYoMM International Symposium On Theory and Practice of Robots and Manipulators, Udine, Italy, September, 357–375.

Hodge, J. (1984) "The Space Station Program Plan." *Aerospace America,* September, 56–59.
Hodnette, L. (1974) "Remotely Piloted Vehicles." *National Defense,* March–April, 422–424.
Hoeber, F. (1977) *Slow To Take Offense: Bomber, Cruise Missiles, and Prudent Deterrence.* Washington, DC: Georgetown University's Center for Strategic and International Studies.
Hoffman, R. (1983) "The Mini-RPV: Tomorrow's Future Multiplier?" *Unmanned Systems*, Spring, 23–26.
Hogg, I. and Weeks, J. (1980) *The Illustrated Encyclopedia of Military Vehicles.* Englewood Cliffs, NJ: Prentice-Hall.
Holliday, D. (1984) "FASTARM." *National Defense,* February, 30–34.
Hudgins, D. (1985) "ROBAT Prototypes Are Taking Shape Here." *TACOM Report,* May 30, 4–5.
"Hughes Lab is Focus of AI Work." (1986) *Aviation Week & Space Technology*, February 17, 87–91.
Humphries, O. (1982) "Star Wars Is Already Here!" *Airpower,* May, 34–51.
Hunt, V. Daniel (1986) "Should Machines Make the Decisions?" *Washington Technology,* December 11, 10.
"Hynes Drone Helicopter For U.S. Army." (1985) *Jane's Defence Weekly,* February, 246.
Isler, W. (1986) "Developments With DARPA's ALV." *Unmanned Systems,* Spring, 22–23.
Jackson, J. (1984) "A Blimp By Any Other Name." *Defense Science 2002+*, June, 65–69.
Jackson, J. (1983) "Airships for the 1990s?" *Naval War College Review,* January–February, 50–60.
Johnston, J. (1984) "Future Prospects For Underwater Robotics." *Unmanned Systems*, Summer, 17–20.
Kannmacher, G. (1969) Walking Machine. In *A Survey of Major Activities of USATACOM Laboratories* (1 July 1967–30 June 1968), Vol. I, 26–27. Washington, DC: U.S. Government Printing Office.
Kennedy, F. (1984) "Sea Services: Unmanned Vehicles at Sea." *National Defense,* November, 12–14.
Kennedy, F. (1986) "U.S. Naval Aircraft and Missile Development in 1985." *Proceedings* (Naval Review), 68–75, 321–325.
Key, W. (1984) "20-Meter Coastal Minehunter With Mine Sonar and Remote Control Vehicle Capabilities." Paper presented at the International Symposium on Mine Warfare Vessels and Systems, London, June.
Klaar, W. (1977) "Drone and RPV Programmes at Dornier." *Dornier Post,* February, 17–21.
Klaar, W. (1975) "RPVs at Dornier—A Status Report." *Dornier Post*, December, 16–17.
Klass, P. (1985) "DARPA Envisions New Generation of Machine Intelligence Technology." *Aviation Week & Space Technology,* April 22, 46–54.
Kolcum, E. (1986) "Search Team Focuses Efforts on Retrieving Orbiter Wreckage." *Aviation Week & Space Technology*, February 10, 59–61.
Kovit, B. (1985) "The Manned Space Station." *Horizons*, Vol. 21, Number 1, 8–18.
Kozlov, N. and Balanin, Ye. (1983) "Military Robots." *Soviet Military Review,* April, 26–27.

LaValley, R. (1985) "RPV Instrumentation Then and Now." *Unmanned Systems*, Fall, 20–23, 30, 31.

Larson, D. (1983) "Unmanned Vehicle Systems: A Key Addition to Our Tool Kit." *Unmanned Systems*, Spring, 10–13.

Laumer, K. (1986) *Rogue Bolo*. New York: Baen Books.

"Lawrence B. Sperry—Aviation Pioneer" (Part 1). (1985) *Sperry Voice*, October/November, 4.

"Lawrence B. Sperry—Aviation Pioneer" (Part 2). (1985) *Sperry Voice*, December, 3.

Ledeen, M. (1979) "Future Weapons." *Penthouse*, February, 117–121, 166, 167.

Lee, W. and Vuk, M. (1984) "A Mobile Military/Security System: Tactical Application of the PROWLER Robot." *Unmanned Systems*, Spring, 28–31.

Lindauer, B., Fini, J. and Hill, J. (1985) "Military Robotics: An Overview." *Robotics Age*, November, 16–21.

Liston, R. and Mosher, R. (1968) "A Versatile Walking Truck." Paper presented at the 1968 Transportation Engineering Conference (ASME–NYAS), Washington, DC.

"Lockheed Conducts Extensive Research on Aquila Variations." (1986) *Aviation Week & Space Technology*, April 28, 101.

"Lockheed Experiments With Unmanned Parafoils." (1985) *Defense News*, November 18, 18.

Lopo, J. (1984) "Tactical Autonomous Weapon Systems." *Unmanned Systems*, Spring, 7–9.

"LTA Technology Assessment." (1983) Paper presented at the AIAA Annual Meeting and Technical Display, Long Beach, CA.

Lynch, R. and Nugent, M. (1986) "Military Applications of Robotics: The USAARMS Approach." *Armor*, May/June, 45–48.

Manners, G. (1985) "Scicon's Intelligent Underwater Robot." *Jane's Defence Weekly*, September 21.

"Manufacturers Tailor Basic Engines For RPV Missions." (1986) *Aviation Week & Space Technology*, April 28, 111.

Marsh, P. (1985) *Robots*. London: Salamander Books.

Maurice, P. (1986) "British Research Center Concocts Demonstrator for Remote Tasks." *Defense News*, May 5, 14.

Maurice, P. (1986) "Robotic Aircraft SOARFLY Likely to Steal Show at British Army Equipment Exposition." *Defense News*, June 23, 3.

"MBB's Drone Family Concept For A Wide Range of Applications." (1985) *MBB International*, June, 8.

McGhee, R. and Waldron, K. (1985) "The Adaptive Suspension Vehicle Project." *Unmanned Systems*, Summer, 34–43.

Melman, Y. (1985) "Skyeye Crash May Help Israel Sales." *Jane's Defence Weekly*, September, 7.

Melzer, W. (1973) "Aerodyne Testing." *Dornier Post*, January, 17–20.

Melzer, W. (1974) "Continuing The Aerodyne Experimental Study." *Dornier Post*, February, 18–21.

Metz, W. (1974) "Aerial Targets." *National Defense*, March/April, 445–447.

Meystel, A. (1983) "IMAS: Evolution of the Unmanned Vehicle Systems." *Unmanned Systems*, Fall, 12–18.

Miller, W. (1979) "Semi-Buoyant Systems Research at Aereon Corporation." *Aerostation*, Autumn, 4–9.

"Mini-Eyes-In-The-Sky." (1984) *Defense Attache*, May, 37.

"Mini-RPV Systems for the USMC: Needed Now–Ready Now." (1984) *Unmanned Systems,* Fall, 30–35.
Mirick, C. (1946) "A Wild-Goose Chase: Early Navy Work on Pilotless Aircraft and Ships." *Proceedings,* July, 947–951.
"Mobile Robot for Perimeter Patrol." (1984) *Defense Electronics,* September, 188.
Morrocco, J. (1986) "Air Force Moves Towards New RPV System." *Defense News,* March 3, 1, 14.
Morrocco, J. (1986) "Air Force Targets Technologies, Weapon Systems for the Future." *Defense News,* February 24, 4, 29.
Morrocco, J. (1986) "Arms Studying Use of Dogfighter RPVs." *Defense News,* April 21, 15, 17.
Morrocco, J. (1986) "Lockheed Makes Last-Ditch Effort to Keep Aquila In The Air." *Defense News,* January 27, 3.
Morrocco, J. (1986) "Military Designing Kamikaze Drones." *Defense News,* April 4, 1, 10.
Morrocco, J. (1986) "Munitions Movers Readying Master Plan for Smart Weapons." *Defense News,* July 7, 10–11.
Morrocco, J. (1986) "Northrop, IBM Joining Forces to Bid for Navy RPV Contracts." *Defense News,* April 21, 10–11.
Morrocco, J. (1986) "Son Of Stealth, Among Air Force Priorities for the Next Century." *Defense News,* May 26, 18.
Mourey, D. (1982) "RPV Airframe Trends." *National Defense,* September, 22–25.
"Moving Land Target Systems." (1978) *Remotely Piloted Magazine,* January, 14, 27.
Munson, K. (1985) "RPVs—Who are the Real Remote Pilots?" (Part 1). *Jane's Defence Weekly,* August 24, 360–364.
Munson, K. (1985) "RPVs—Who are the Real Remote Pilots?" (Part 2). *Jane's Defense Weekly,* August 31, 411–413.
"NASA Assesses External Tank's Role in Challenger Accident." (1986) *Aviation Week & Space Technology,* February 17, 18–20.
"NATO Strategies, Limited Resources Increase Uses for Unmanned Systems." (1986) *Aviation Week & Space Technology,* April 28, 51–56.
"Navigation Challenges Autonomous Vehicle." (1985) *Aviation Week & Space Technology,* April 22, 74.
"Navy Jumps For A New Drone." (1985) *Navy News & Undersea Technology,* July 19, 6.
"Navy Using Wing Manufacturing's Cruise 3 RPV as Sensor Testbed." (1986) *Aviation Week & Space Technology,* April 28, 121.
"New ROVs For Navies." (1984) *Defence Attache,* May, 5.
"New Software Allows Manual Repositioning of RPV Laser Spots." (1986) *Aviation Week & Space Technology,* April 28, 101.
Newman, N. (1972) "Remotely Manned Systems: Technology's Helping Hands." *Naval Institute Proceedings,* 115–116.
"Northrop Delivers RPVs to Navy to Aid in Development Proposal." (1986) *Aviation Week & Space Technology,* April 28, 123–127.
Office of Robotics and Autonomous Systems. (1984). "Naval Sea Systems Command Integrated Robotics Program." *Annual Report*, Fiscal Year 1984, (SEA 90G), December. Washington, DC: Government Printing Office.
O'Neil, R. (1984) *Suicide Squads.* New York: Ballantine.
"Operational Requirements Drive Procurement of RPVs." (1986) *Aviation Week & Space Technology,* April 28, 42–48.

Ortelli, A. and Muia, C. (1984) "Mine Identification and Neutralization System Developed by the SMIN Corporation." Paper presented at the International Symposium on Mine Warfare Vessels and Systems, London.

Ottaway, D. (1982, June 14) "Israel Said to Master New Technology to Trick and Destroy Soviet-Made Rockets." *The Washington Post,* pp. A1, A25.

"Pacific Aerosystems Continues Development of Heron 26 Vehicle." (1986) *Aviation Week & Space Technology,* April 28, 63–66.

"PAP-104 Mine Destruction Vehicle." (1983) *Armed Forces Journal,* December, 60.

Parker, P. (1985) "Progress Report on the RPV." *Marine Corps Gazette,* May, 38–40.

"Partners Agree to Production of CL-289 Drone." (1986) *Aviation Week & Space Technology,* March 3, 93–97.

Perrin, J. (1986, January 26) "The Robots Are Coming!" *The Washington Post Book World*, p. 7.

Persson, L. (1984) "An Acquisition Model for a Small Country." *Unmanned Systems,* Fall, 41–43.

Pincus, W. and Auerbach, S. (1986, January 30) "Loss of Challenger Renews Arguments for Using More Unmanned Spacecraft." *The Washington Post,* p. A18.

Pincus, W. (1986, July 15) "More Drone Tests Sought in Honduras." *The Washington Post*, p. A5.

"Pinguin Flies In Water." (1985) *MBB International*, June, 9.

Pole, K. (1983) "Airships: Key to Canada's Frontiers?" *Canadian Geographic,* February/March, 10–16.

Pournelle, J. (1983) *There Will Be War.* New York: TOR Books.

Powell, K., Cohen, A. and Craig, M. (1984) "Space Station Design: Innovation and Compromise." *Aerospace America,* 70–72.

"Powerplant Advances Tied to New Technologies." (1986) *Aviation Week & Space Technology,* April 28, 111–112.

"Project Forecast II Completed." (1986) *Air Force Magazine,* April, 22–23.

"Prowler 60 Robotic Patrol Vehicle." (1984) *International Defense Review,* November, 1757.

Purdy, K. (1985) "Major Role for Robots in Defense and Aerospace." *Jane's Defence Weekly,* April, 616.

Rae, M. (1984) "A Sensor Manufacturer's Experience of Unmanned Autogyros Development." *Unmanned Systems,* Fall, 12–17.

"Red Sea Mine–New Soviet Type." (1984) *Jane's Defence Weekly,* October, 6.

Reed, M. (1984) "The National Dynamics Rhomboid Wing Eyrie RPV System." *Unmanned Systems*, Fall, 24–29.

"Reusable Shuttle Stage in Offing at Astrotech Corp." (1986) *Defense News,* February 3, 12.

Robinson, D. (1982) "The Aereon Corporation and Its Lifting Body Airships." *Aerostation,* Summer, 25–30.

Robinson, R. (1985) "Unmanned Untethered Submersibles for National Defense." *Unmanned Systems,* Fall, 32–37.

Rogers, B. (1986) "NATO's Conventional Defense Improvement Initiative: A New Approach to an Old Challenge." *NATO's Sixteen Nations,* July, 14–27.

Rogers, M. (1984) "Birth of the Killer Robots." *Newsweek,* June 25, 51.

Rose, K. (1986) "Robots That Go All The Way." *Army,* January, 18–27.

Ross, G. (1985) "Robbie the Robot." *Airman Magazine,* August, 3.

"Rotec Using Ultralight Design to Develop RPV Platform." (1986) *Aviation Week & Space Technology,* April 28, 116, 121.

"ROV for Dredging Work." (1985) *Jane's Defence Weekly,* August 17, 333.

"RPV Application Values." (1985) *The Achiever*, June, 6.

Russell, D. (1983) "Israel RPVs: The Proven Weapons System DoD Will Not Buy." *Defense Electronics,* March, 86–92.

Russell, M. (1983) "ODEX 1—The First Functionoid." *Unmanned Systems*, Fall, 9–11.

Schiller, R. (1982) "Propulsion Trends for Unmanned Air Vehicles." *National Defense,* September, 26–30.

Schmidt, W. (1986, February 20) "Debris from Shuttle Booster is Identified." *The New York Times,* p. 15.

Schultz, J. (1984) "Cruise Missile Deployment Marked By System Upgrades and Operational Tests." *Defense Electronics,* May, 47–64.

Schwartz, M. (Ed.) (1982) *The Encyclopedia of Beaches and Coastal Environments.* Stroudsburg, PA: Hutchinson Ross Publishing Co.

"Sea Eagle and Pluto: Mine-Disposal Submersibles from Sweden and Italy." (1984) *International Defense Review,* April, 500.

"Search Continues For Shuttle Debris Off Coast of Florida." (1986) *Aviation Week & Space Technology,* February 3, 22–23.

Seemann, G. (1983) "Unmanned Vehicle Systems Use Ranges from Pure Science to Military Tactics." *Defense Systems Review,* July, 44–45.

Shaker, S. (1985) "New Airship Designs." *National Defense,* July/August, 66.

Shaker, S. (1986) "Remotely Operated Military Unmanned Submersibles." *National Defense,* May/June, 61–66.

Shaker, S. (1986) "Space Robotics Systems and Technology." *National Defense,* April, 33–36.

Shaker, S. (1986) "Unmanned Lighter-Than-Air Vehicles." *Unmanned Systems,* Fall, 24–29.

Shaker, S. (1985) "Unmanned Vehicle for Logistic Missions." *National Defense,* December, 38–44.

Shaker, S. and Shaker, H. (1986) "Israeli Weapons Technology and the U.S. Military." *National Defense,* March, 38–41.

Shaker, S. and Wise, A. (1985) "Research Progress in Unmanned Vehicles." *Armor,* July/August, 33–36.

Shaker, S. and Wise, R. (1986) "Walking to War: Research Efforts in Legged Vehicles." *National Defense,* March, 59–62.

Simpkon, R. (1985) *Race to the Swift: Thoughts on Twenty-First Century Warfare.* London: Brassey's Defence Publishers.

Simpson, T. and Dekuk, R. (1984) "Into the Wild Black Yonder." *Space World,* April, 9–14.

Skutch, L. (1985) "Robotics—Only the Beginning." *Military Logistics Forum,* May, 71–73.

Skyeye. (1985) In *Department of Defense Appropriations For 1986, Hearings before Subcommittee of the Committee on Appropriations, House of Representatives 99th Congress, First Session,* 908–912. Washington, DC: U.S. Government Printing Office.

Sokol, A. (1944) "German Experiment With Remote Control During the Last War." *Naval Institute Proceedings,* February, 187–189.

"Soviets Captured Target Drone Claim." (1984) *Jane's Defence Weekly,* June 16, 942.

"Soviet Intelligence Reconnaissance Drones Developed to Meet Expanding Soviet Needs." (1985) *Jane's Defence Weekly,* August 10, 260–261.

Stansell, K. (1985) "Monitoring the Enemy Position." *Marine Corps Gazette,* May, 40–43.

Stansell, K. (1986) "The Skyeye RPV: An NDI Solution for Today's Unmanned Aerial Vehicle Requirements." *Unmanned Systems,* Spring, 10–11, 30.

Stern, K. (1985) "Year 2010 Fighter May Employ Electronic Backseat Crewmember." *Aviation Week & Space Technology,* June 24, 99–102.

Stewart, J. (1986, January 11) "Robots Could Replace Humans on the Battlefield of the Future." *The Atlanta Journal/The Atlanta Constitution Weekend,* pp. 1A, 19A.

Swann, C. (1985) "Robot Mini-Subs Replace Pro Divers." *Popular Science,* November, 80–81.

Sweeney, J. (1970) *A Pictorial History Of Oceanographic Submersibles.* New York: Crown.

Taylor, J. (Ed.) (1977) *Jane's Pocket Book of Remotely Piloted Vehicles.* New York: Macmillan.

Taylor, J. and Patterson, W. (1974) "The Navy Looks Aloft." *National Defense,* March/April, 425.

"Teledyne Ryan Curtails RPV Development Programs." (1976) *Aviation Week & Space Technology,* April 28, 66.

"Telerobotics May Be in Corps Future." (1986) *Marine Corps Gazette,* February, 7.

"Thanks For The Headaches." (1985) *International Combat Arms*, March, 9.

Theile, B. (1984) "The CL 289 Reconnaissance Drone System." *Unmanned Systems,* Fall, 44–46.

Todd, D. (1983) *Walking Machines: An Introduction to Legged Robots.* New York: Chapman and Hall.

"Tomahawk Programmable Warhead Explodes Above Target, Destroys It." (1986) *Defense News,* April 28, 28.

Townsend, J. (1985) "Future Military Applications of Unmanned Systems." *Unmanned Systems,* Spring, 21–30.

Townsend, J. (1985) "Unmanned Systems for NATO." *NATO's 16 Nations,* August, 72–77.

Townsend, J. (1985) "Unmanned Systems for NATO." *Unmanned Systems,* August, 72–75.

Tucker, J. (1986) "Submersibles Reach New Depths." *High Technology,* February, 17–24.

Ulsamer, E. (1977) "Flying Robots with a Lethal Sting." *Air Force Magazine,* February, 29–34.

United States Department of Commerce. (1979) *Remotely Operated Vehicles.* Washington, DC: U.S. Government Printing Office.

"Unmanned Crop-Sprayer Flies." (1985) *Flight International,* December 21–28, 14.

"U.S. Army Plans To Evaluate Canadair CL-227 Sentinel." (1986) *Aviation Week & Space Technology,* April 28, 103–105.

U.S. Congress, Office of Technology Assessment. (1985) *Ballistic Missile Defense Technology* (OTA-1SC-254). Washington, DC: U.S. Government Printing Office.

Vandersteen, T. (1985) "RPV Battlefield Payloads." *Journal of Electronic Defense*, October, 137–148.

Wald, D. (1985) "Artificial Intelligence in 2000 A.D." *Defense Electronics,* July, 111–113.

"Walking Machine." (1967) *United States Army Tank–Automotive Command Historical Summary, Fiscal Year 1987* (Vol. I) 307–308. Washington, DC: U.S. Government Printing Office.

Walters, B. (1985) "Time for Inflatables to Take Off." *Jane's Defence Weekly,* October 19, 883–884.

Weinberger, C. (1987). "Importance of Conventional Defence." *Jane's Defence Weekly,* January 10, 13.

"Weinberger Sees Threat to NATO." (1987, January 8) *The Washington Times,* p. 1A.

"Westinghouse Applies Expert System Techniques to Trouble-Shooting Tasks." (1986) *Aviation Week & Space Technology,* February 17, 91.

Westneat, A., Blidbery, D. and Corell, R. (1983) "Advances in Unmanned, Untethered Underwater Vehicles." *Unmanned Systems,* Winter, 8–13.

Williams, D. (1986) "Air Force Targets Artificial Intelligence for Development of 'Brilliant' Weapons." *Defense News,* July 7, 8.

Williams, D. (1986) "Artificial Intelligence Emerges from Laboratory." *Defense News,* May 12, 15.

Williams, G. (1986, March 3) "Mirror-Reflected Laser Suggested to Shield Allies." *The Washington Post,* p. A7.

Wilson, G. (1986, October 21) "Budget Cuts: SDI's Priority Hurt Conventional Military Readiness." *The Washington Post,* p. 10.

Wilson, G. (1985, September 5) "Unmanned Weapons Gain Backing: Official Stresses Saving Lives, Money." *The Washington Post,* p. A15.

Wilson, G. (1982, June 14) "U.S. Navy, Guided by Eyes in the Sky, Launches a New High-Tech Harpoon." *The Washington Post,* pp. A1–A2.

Index

(Italicized page numbers refer to photo pages)

About the Authors/Editor

Steven M. Shaker is a senior staff member for advanced planning with the BDM Corporation and an authority on robotic weapon systems. He has published numerous articles on this subject and on other advanced weapon concepts. He is a frequent contributor to *National Defense* magazine and has also been published in *Armor*, *Defense Management Journal*, *International Combat Arms*, *The Marine Corps Gazette*, *NBC Defense and Technology*, the *Journal of Defense & Diplomacy*, and *Unmanned Systems*. Mr. Shaker is a former senior analyst on future weapon systems with the Naval Air Systems Command. He is an adjunct professor with the Florida Institute of Technology's graduate program in management science, specializing in defense acquisition management.

Alan R. Wise is a developmental engineer concentrating in manned and unmanned off-road vehicles. He has contributed to the design of many advanced suspension systems for all-terrain vehicles. Mr. Wise is also a writer and photojournalist on robotic vehicles and weapon systems, as well as motor racing technologies. He is a frequent contributer to *National Defense*, *Armor*, *MOTOR* magazine, and the *Journal of Defense & Diplomacy*.

FOREWORD AUTHOR

Robert Finkelstein is president of Robotic Technology, Inc., a company that provides services and systems in the field of combat robotics. He is also president of the Association for Unmanned Vehicle Systems Capital Chapter and serves on the AUVS board of trustees. Mr. Finkelstein writes a regular column for the association's journal, *Unmanned Systems*.

GENERAL EDITOR

Perry M. Smith, retired USAF major general, is former director of plans, USAF, and former commandant of the National War College, Washington D.C.